CAPITALISM WITH MORALITY

CAPITALISM WITH MORALITY

D. W. HASLETT

CLARENDON PRESS · OXFORD
1994

Oxford University Press, Walton Street, Oxford OX2 6DP
Oxford New York Toronto
Delhi Bombay Calcutta Madras Karachi
Kuala Lumpur Singapore Hong Kong Tokyo
Nairobi Dar es Salaam Cape Town
Melbourne Auckland Madrid
and associated companies in
Berlin Ibadan

Oxford is a trade mark of Oxford University Press

Published in the United States
by Oxford University Press Inc., New York

British Library Cataloguing in Publication Data
Data available

Library of Congress Cataloging in Publication Data
Haslett, D. W.
Capitalism with morality/D. W. Haslett.
Includes bibliographical references and index.
1. Economics—Moral and ethical aspects. 2. Capitalism—Moral and
ethical aspects. 3. Comparative economics—Moral and ethical
aspects. I. Title.
HB72.H34 1994 330—dc20 93–44416
ISBN 0–19–828553–1

1 3 5 7 9 10 8 6 4 2

Typeset by Cambrian Typesetters Frimley, Surrey
Printed in Great Britain
on acid-free paper by
Bookcraft Ltd., Midsomer Norton, Nr Bath, Avon

To my children
Heidi and Erik,
with love

PREFACE

NOT long ago, Robert Heilbroner wrote: 'Less than seventy-five years after it officially began, the contest between capitalism and socialism is over: capitalism has won' (1989: 98). The 'contest' may be over. Yet the deep feelings of discontent that many have with capitalism—the very feelings that gave rise to socialism in the first place—remain as strong as ever. This discontent stems not primarily from any dissatisfaction with capitalism's overall productivity. It stems instead from the vast inequalities in wealth and opportunities that seem to characterize capitalism and from the hopelessness and alienation of many ordinary working people. In other words, this discontent stems from what are perceived to be certain moral deficiencies of capitalism. I shall argue that these preceptions are justified, that current capitalism is indeed morally deficient. This book is about whether, without abandoning capitalism, these serious deficiencies can be overcome.

The debate over capitalism versus alternative systems is, in the hands of academic economists, conducted these days with unprecedented precision and eloquence. This book, however, seeks to broaden this debate in three ways. First, it seeks to make the basic arguments from the field of economics accessible to a broader audience than just economists and their students. Instead of these arguments being put here in terms of equations and graphs, as they usually are by economists, they are put in plain English. Broader participation in the debate over alternative economic systems is essential. For, in a democracy, which system is most justified will, ultimately, be decided not by these academic economists, but by democratic vote.

The second way in which this book seeks to broaden the debate is by evaluating economic systems in terms of a broader range of criteria than is usually used by academic economists. Economists usually evaluate economic systems in terms of standard economic criteria, such as productivity, growth, and full employment. This book, however, uses moral criteria as well, such as freedom, justice, and equal opportunity.

The third way in which this book seeks to broaden the debate over alternative economic systems is by introducing another alternative into the debate: an alternative that, I shall argue, does, in fact, avoid the moral deficiencies of current capitalism without, however, abandoning capitalism itself and its traditional strengths.

Evaluating economic systems in terms of moral criteria presupposes, of course, a moral perspective from which this evaluation can proceed. The first chapter of this book, therefore, is devoted to setting out and defending just such a perspective. Most books begin with the chapter that is easiest to read and progress slowly, step by step, towards more difficult chapters. This book, however, begins with the most difficult chapter, namely, this chapter on moral theory. It is the most difficult not because it presupposes any prior knowledge, but simply because the topic is the most abstract.

But this chapter on moral theory, as abstract as it may be, is nevertheless necessary. Do not misunderstand me. It is necessary not because, from the methodology of moral justification set out there, I proceed to justify all the moral norms and values that I use later in the book. There already is, I think, fairly widespread agreement that these particular moral norms and values are justified (even though there may not be widespread agreement about the details of their analysis). And, so as to keep this book both a reasonable length and within my capacity to write, my policy in writing it was as follows: if there already is widespread agreement that a moral norm or value is justified, then do not attempt to justify it.

Why then is Chapter 1 necessary? First of all, a book that purports to evaluate economic systems from a moral perspective certainly ought, in fairness to its readers, to make that perspective as clear as possible from the start. But Chapter 1 is necessary for still another reason. In this chapter I try to come to terms with those who, ultimately, are my main opponents: sceptics.

Sceptics often argue as follows. Conclusions about matters of morality can never be proven beyond any doubt; all such conclusions, therefore, are, in the end, arbitrary. These sceptics, however, are mistaken. To avoid being arbitrary, conclusions about matters of morality need not be proven beyond any doubt. They need only be supported by careful argumentation—provided, however, that we at least have a defensible methodology that delineates what *counts* as proof. It is just such a methodology that I try to make clear and defend in Chapter 1 (along with Sect. 2.2).

Although no less than the entire chapter is necessary for a full understanding of this methodology and the argument against scepticism, by reading just Sections 1.1, 1.2, 1.3, and 1.11, a simplified understanding can be gained by readers who, for one reason or another, do not require any more than that.

The main conclusions I reach in the next two chapters (Chs. 2 and 3) are, very briefly, as follows. Neither libertarian capitalism nor central-planning socialism is morally defensible. Nor, however, is capitalism as it exists today. Nevertheless, any morally defensible economic system must, for the sake of freedom and productivity, be built around a free market and the private ownership of capital. But it must also be a system that is far more successful than any current forms of capitalism in providing equal access to basic necessities, equal opportunity, and worker autonomy.

I try to defend such a system—a capitalism with morality—in the final three chapters of this book. This system features, among other things, elements of worker control, an inheritance quota, and a redistribution of income that does not compromise either freedom or productivity.

Although each of the chapters of this book is closely related to the others, each can be understood independently. What I mean is that, although each of the chapters presupposes the truth of the others, it does not presuppose a knowledge of the others.

Sections 1.7–1.10 of Chapter 1 make use of excerpts from an article that appeared in *Economics and Philosophy*, entitled 'What is Utility?' (1990); and Chapter 6, except for the final two sections, is a revised version of an article that appeared in *Philosophy & Public Affairs*, entitled 'Is Inheritance Justified?' (1986). I thank the editors and publishers of these journals for their permission to use this material.

Each of the following people has read, either in typescript or as it appeared previously in print, one or another of the parts of this book, and has been kind enough to offer many constructive comments: Christopher Boorse, John Broom, Tracy Evans, Milton Friedman, John C. Harsanyi, Saul D. Hoffman, Daniel M. Housman, Jeffrey A. Jordan, Bill E. Lawson, Bertram Levin, Michael S. McPherson, Edgar Page, Chris Robe, Laurence S. Seidman, Roy Smith, L. W. Sumner, and James Thornton. I have also received a number of constructive comments from many of my students, especially those in my graduate course on business ethics. Although those who have offered comments may not always agree with my responses, their comments have been most helpful.

I wish to thank the University of Delaware for having provided me with a sabbatical during which part of this book was written, and Mary Imperatore for her help in preparing the typescript.

Finally, I wish to express a special thanks, as always, to my wife, Beth.

D. W. H.

Newark, Delaware
June 1993

CONTENTS

Contents

1

IS THERE A CORRECT ANSWER?

The critics of capitalism see the vast differences in wealth and opportunities between rich and poor that this system breeds, they see the unfair
burdens that it places upon ordinary working people, and they conclude
that capitalism cannot be justified. These vast inequalities and unfair
burdens are indeed grave, moral deficiencies of capitalism, as it exists
today. Can, however, these deficiencies be overcome, without abandoning
capitalism and its traditional strengths? Or, as the critics contend, should
we turn instead to some form of socialism; if not to central-planning
socialism, which has fallen out of favour recently, then perhaps to market,
or worker-control socialism? These are the questions to which this book is
addressed.

Economic systems—capitalism, socialism, worker control, and so on—
are, of course, systems for producing goods and services. So which system
is most justified depends in part simply upon which is capable of turning
out the most goods and services. But which is most justified depends upon
much more than merely this. Economic systems also delineate, in general
terms, how goods and services are to be distributed throughout society. So
even though a system turns out a greater quantity of goods and services
than any other, it may not be the one that is most justified if it does not
distribute these goods and services fairly. Moreover, it may not be the one
that is most justified if it does not allow for as much freedom, or as much
equal opportunity, as alternative systems. In short, to determine which
economic system is most justified we must take into account not only
technical, quantitative matters, but matters of morality and values as well,
matters such as justice, freedom, and equal opportunity. Since the
technical, quantitative matters relevant for choosing among alternative
economic systems have been examined at great length and with much skill
by economists, I shall not concentrate primarily upon these matters here.
Rather, I shall take up where economists generally leave off by
concentrating primarily upon matters of morality and value. I make no
attempt to adhere to the dictates of any ideology, be it Marxism,
libertarianism, liberalism, or conservatism, but attempt to adhere only to
the dictates of morality.

The system that I shall conclude is, all things considered, most justified,

and that I shall set out in the final three chapters of this book, will be
neither capitalism nor socialism in any of their current forms. It is a system
that, I shall argue, will bring about more equality of opportunity and less
poverty than any modern, industrialized society has yet realized, and will
do so without any overall loss in personal freedom or economic
productivity. Since this system differs significantly from any current
system, my conclusions will no doubt appear far too liberal for many
people today. But since this system does not include any minimum wages,
labour unions, progressive taxation, government ownership of capital
goods, or government planning of the economy, it is certainly not
liberalism of any traditional sort.

1.2 THE SCEPTICS

But how, it may be asked, can conclusions such as these—conclusions that
depend upon value judgements—ever be shown to be well supported? This
is a matter about which there is much controversy and misunderstanding.
Many view value judgements merely as expressions of personal whim, or
expressions of self-serving class ideology. These are the sceptics.

This scepticism is often found among economists. Take, for example,
Robert L. Heilbroner and Lester C. Thurow. In their introductory
textbook, *Understanding Microeconomics*, they write as follows: '. . . you
may take the position that the best distribution of income entrusts one man
with virtually all the nation's wealth, while the rest of the people are his
slaves. Perhaps you will say this accords with your judgment of what is
best. There is no way of "proving" your values wrong' (1975: 269). 'There
is no correct solution for . . . any . . . problem involving value judgments'
(ibid. 271). Similarly, Harry D. Hutchinson, in his introductory economics
textbook, writes: 'Is it just—fair—to pay people solely according to what
their productive contribution is "worth", in the market sense? This
question, involving (as it does) personal value judgment, has no right or
wrong answer' (1973: 85–6). When it comes to questions involving value
judgements and, in particular, ones involving the meaning of justice,
Heilbroner and Thurow tell us that economists 'have politely bowed out of
the discussion', leaving such questions to philosophers (1975: 267).

When evaluating alternatives to the status quo is unavoidable, many
mainstream economists these days compensate for their scepticism by
appealing to the so-called 'Pareto' criterion. According to this criterion,
some alternative *A* is superior to (or more 'efficient' than) some other
alternative *B* if and only if at least one person is better off with *A* than *B*,
and no one is worse off. 'Better off' and 'worse off' are to be understood

here to be merely matters of preference; a person is better off with *A* rather than *B* if and only if this person prefers *A* to *B*.

The Pareto criterion is indeed not without its legitimate uses. But for evaluating alternatives to the status quo, especially alternatives as complex as entire economic systems, it is not an adequate substitute for moral values and norms. To begin with, an interpretation of 'better off' and 'worse off' merely in terms of preferences is, as I shall try to show later in this chapter, inadequate. But the more obvious objection to the Pareto criterion is this: no alternative as complex as a proposed economic system can possibly avoid making at least some people worse off; thus the Pareto criterion rules out every such alternative to the status quo from the start. No criterion could be more protective of the status quo, no matter how evil, or of entrenched special privileges, no matter how arbitrary, than this.

This sort of objection to the Pareto criterion can be avoided, it has been argued, if the criterion is qualified by the so-called 'compensation test', which works as follows. If those who will be made better off by some proposal could still come out ahead even after paying those made worse off an amount of money equal to their loss, then the proposal is 'Pareto' superior to the status quo after all.

The compensation test prevents the Pareto criterion from ruling out every alternative to the status quo from the start. But it is far from sufficient for vindicating the use of the Pareto criterion for evaluation in general. First of all, large-scale proposals are often such that monetary compensation for those made worse off would not be possible even in principle; full monetary compensation would destroy the proposal. Take, for example, a proposal to abandon capitalism in favour of socialism. By paying off all capitalists—and their heirs—an amount equivalent to their losses, the very privileges and inequalities that socialism was designed to eliminate might then be forever entrenched. And if there were no intention of ever actually paying any such compensation, the Pareto criterion would still be insufficient, by itself, for evaluating the proposed change, for then other norms and values would be needed for determining the morality of no compensation.

Moreover, say the proposal would make certain people worse off by taking from them certain rights, such as the right to use their property as they see fit, the right to free speech, the right to the religion of their choice—or were to make them worse off by undermining certain abstract things they value, such as autonomy or equal opportunity. How could these people, or any one else, possibly determine how much money would be adequate compensation for these sorts of losses? How to do this is especially problematic if, as I try to show in Sections 1.7 and 1.8, we cannot accurately measure people's gains and losses merely by reference to their current preferences anyway. Indeed, what if these people were made worse off through having their life expectancies greatly shortened? Could any amount of money be adequate compensation for that? There are, in short, many things of value that should be taken into account in evaluating complex proposals, but that resist any satisfactory translation into mere cash. For these various reasons, the Pareto criterion is far from compensating for scepticism about value judgements, and it will not play a major role in this enquiry.

Scepticism about value judgements is not, of course, limited to economists (and not all economists are sceptics). It is common among other social scientists, philosophers, lawyers, intellectuals, business executives, college students, and many others. Since scepticism about value judgements is in fact so widespread, before attempting any value judgements here I shall, in this chapter, try to show how they can, legitimately, be made. The abstract and highly controversial matters of moral theory I consider in this chapter may appear far removed from the specific evaluations of economic systems that follow. But evaluative enquiries inevitably raise these matters of moral theory sooner or later. Moreover, one's theoretical perspective will inevitably influence one's specific conclusions. So, for the sake of clarity, let me, from the start, reveal and try to defend my own theoretical perspective.

First, and most importantly, those who claim that questions about what is valuable and morally right have no correct answers are, I believe, making an enormous mistake; they do have correct answers. If I thought they had no correct answers, then not only would I politely bow out of any *discussions* about what was valuable and morally right, as Heilbroner and Thurow tell us that economists do, but I would also bow out of trying to *live* according to what was valuable and morally right. For if questions about what was valuable and morally right had no correct answers, then trying to live according to what was valuable and morally right would make about as much sense as trying to reach the end of the rainbow.

Consider, for example, the question: 'Was it morally wrong for Hitler to have killed six million innocent people in concentration camps?' Do Heilbroner and Thurow really believe this question has no correct answer? I hope not. The answer—the *correct* answer—is: 'Yes, it was morally wrong.' And if this value question about what is morally right has a correct answer, then surely it is reasonable to suppose that other such questions do also. The correct answers will by no means always be obvious. And the correct answers may well depend upon the particular circumstances of the society or individuals in question—that is, the correct answers may well be relative to circumstances. But correct answers there are.

I do not know what leads so many people to believe otherwise. Some people may perhaps believe that there are no correct answers to questions about what is valuable and morally right—let us call them 'value' questions—because philosophers have been debating for centuries about how we are to go about answering them and have yet to reach any agreement. But this does not mean that no answers are correct. Philosophers have also been debating for centuries about how we can legitimately answer questions about other people's minds—whether other people are in pain, are happy, are sad, and so on. And clearly these questions have correct answers.

To be sure, with value questions one can always play the sceptic's game. It goes something like this. First, any answers to value questions that one's opponent gives are to be challenged. When presented with the grounds for these answers, one is to challenge the grounds. Then the grounds for these grounds, and then the grounds for the grounds for these grounds, and so on, until one's opponent simply throws up his or her hands in despair.

Although this game is easy enough for the sceptic to play, it does not show that value questions have no correct answers. For, as students in introductory philosophy courses soon find out, in being vulnerable to the sceptic's game, value questions are no different from questions of any other sort—including those investigated by economists. Yet, in spite of this game, some questions surely do have correct answers.

Of course, the sceptic may then deny even this. But if the sceptic were thus reduced to denying that any questions at all had any correct answers, then would this not be a plain example of self-refutation through *reductio ad absurdum*? Moreover, what reply might this sceptic give to the question: Do any questions have correct answers? If 'Yes', then this contradicts the sceptic's own thesis that no questions have any correct answers. If 'No', then either that answer is correct or it is not. If it is correct, then it is a counter-example to the sceptic's thesis, thereby proving the thesis to be false; if the answer is not correct, then the sceptic's thesis is, once again, false; either way it is false.

Finally, people are sometimes perversely misled into claiming that value questions have no correct answers by their support for toleration. They seem to think that only if value questions have no correct answers does it make sense for all answers to be tolerated. But the exact opposite is true; only if value questions do have correct answers does toleration make any sense. After all, toleration is itself a value; so without any correct answers to value questions, not even toleration can be rationally supported.

If then value questions do have correct answers, how do we go about finding them? A preliminary answer, correct as far as it goes, is that we do so by appeal to those moral norms or principles, and to those values, that are most justified. But, of course, this answer does not go far enough; it serves only to raise a more fundamental question: Which moral norms and values are most justified? And the sceptics will tell us that this question has, once again, no correct answer. In what follows, I shall try to show that the sceptics are wrong; this question does indeed have a correct answer.

But first we need to get straight about some important distinctions, those between norms, values, and standards. Norms are criteria that prescribe that we act (or not act) in a certain way. Some norms are moral principles, such as 'Do not break your promises' and 'Do not discriminate against anyone on the basis of race, sex or religion'; other norms are rules or principles that have nothing to do with morality, such as 'Brush your teeth every day' and 'Shake hands upon being formally introduced'. All, however, are general prescriptions purporting to distinguish that which is right—morally right, prudentially right, or otherwise right—from that which is wrong.

Values, on the other hand, are criteria purporting to distinguish that which is good from that which is bad. Rather than taking the form of rules, principles, and so on, as do norms, values take the form of virtues, ideals, and so on. Examples of values are 'courage' and 'kindliness' (which are virtues) and 'freedom' and 'equal opportunities' (which are ideals). Values are not as strong as norms; they do not prescribe or prohibit specific acts, as do norms. Rather, they indicate what, generally, is to be striven for, and to be striven for even if incapable of ever being achieved fully. In other words, norms delineate what, morally speaking, we must do, or not do, while values delineate what qualities, morally speaking, are desirable in general, not what we must do, or not do, in any given circumstances.

Finally, inherent in a value will be a standard, a criterion in terms of which things may be *comparatively* evaluated. Take, for example, the value, or ideal, of 'freedom'. And let that which is to be evaluated be social institutions, such as economic systems. The standard inherent in this value is then simply: 'The more freedom a social institution allows, the better (or more justified) it is.'

Whenever, throughout this enquiry, I speak of a 'code of morality', I mean morality in a 'broad' sense which includes not just moral norms, but values and standards as well. Similarly, whenever I speak of 'normative criteria', I mean not just moral norms but values and standards as well.

1.3 TWO PREMISSES

Let me return now to the question at hand: How can moral norms and values (and, of course, the standards inherent in these values) be shown to be justified? I shall begin with basics. Why do we have moral norms and values anyway? What is their point? Why not simply do without them altogether?

Speaking in the most general terms, their point, of course, is our well-being, or welfare. As Thomas Hobbes might say, without moral norms and values, life among a group of human beings would be 'nasty, brutish, and short'. This then is the first premiss upon which my conclusions about the justification of moral norms and values are based: the general point of there being moral norms and values rather than none is that they contribute to human well-being; they, in other words, enable people to 'get on' better than would otherwise be possible. (Note: I shall assume, for the moment, that the concept of personal 'welfare', or 'well-being', is adequately understood. In fact, it is not. But I shall postpone any attempt at clarification until Sect. 1.7.)

This premiss about the general point of there being moral norms and values shall remain undefended here. I do not mean to suggest that this premiss cannot be successfully defended; I think it can. But since I cannot try to defend everything here, certain premisses that would appear to be obviously true (except, perhaps, to philosophers), I shall simply pre-suppose. And no premiss would appear to me to be more obviously true than that the general point of there being norms and values is our well-being. They make things go 'better' for us (cf. Warnock 1971).

But this premiss, as obvious and innocent as it may appear, has enormous ramifications. Anything that has a point (object, purpose) is to be judged according to the extent to which it achieves that point (object, purpose). So if the point of moral norms and values is our well-being, it follows that they are to be judged according to the extent to which they achieve our well-being. Thus we appear, already, to be coming close to an answer to the question: Which moral norms and values are most justified? The answer, in the light of this one premiss, would appear to be: Those most conducive to our well-being.

But then whose well-being, exactly, is 'our' well-being? Is it the well-being of whites rather than blacks, or of males rather than females, or what? In other words, in determining how justified moral norms and values are, whose well-being or interests, exactly, are we to give primary consideration to? This brings me to a second premiss, which is as follows: in determining which moral norms and values are most justified, we are to give equal consideration to *everyone's* well-being—or interests—whether the person be black, white, male, female, or whatever. This premiss means that equal interests are to be taken as equally important, regardless of whose interests they are. (The question of whether even animals should be included within the domain of those to be given equal consideration need not be addressed here, but for arguments that animals should, see Haslett 1987*a*; Singer 1977.) I try to defend the equal consideration premiss at length elsewhere (Haslett 1987*a*). But let me now give at least a partial defence of this premiss here.

Philosophers have long recognized a so-called principle of 'formal' justice that is formulated sometimes as 'Treat like cases alike', and other times as 'Treat all cases alike, unless there is a relevant difference between them'. This is considered a principle of *formal* justice because no substantive conclusions about justice are derivable from it without first specifying exactly what constitutes 'relevant' differences. And, as R. M. Hare has shown (e.g. Hare 1963; 1981), this principle of formal justice is not really a principle of 'justice', or of 'morality', at all, but an application of a broader principle inherent in the logic of ordinary language, which might be put as follows: 'If x is P, and if y is identical to x in all relevant respects, then y is P also.' And, of course, y's being identical to x in all relevant respects means that y has all those properties x has that make it P. Let us call this the principle of 'universalization'.

But now consider this. When there is a disagreement about whether two kinds of things are relevantly different or not, who should have the burden of proof? Say, for instance, that the two kinds of 'things' about which Smith and Jones have this disagreement are white people and black people. Smith claims that black people are different relevantly from white people in a way that justifies our treating blacks as inferiors; Jones claims that there is no relevant difference and that, therefore, blacks and whites should, as the principle of universalization tells us, be treated the same. Who, as between Smith and Jones, should have the burden of proof?

I claim that the burden should always be on whomever is claiming the relevant difference because it is only that person who can meet such a burden. A person claiming the opposite, that there is no relevant difference, can hardly point out no relevant difference. No matter how many alleged relevant differences such a person establishes as non-existent, he still will not have proven that there is no relevant difference, since, in principle, there might always be a relevant difference that he has not yet taken into account. One cannot, in other words, prove a universal, negative claim. But the person claiming there is a relevant difference can, if there really is such a difference, point it out. One can, in other words, prove a particular, affirmative claim. Since, therefore, only the person claiming there is a relevant difference can be in a position to meet a burden of proof, it is only reasonable that the burden be upon him. In the example above, this means that the burden is upon Smith to prove that there is a relevant difference between black and white people that justifies our treating blacks as inferiors, rather than upon Jones to prove that there is no such difference. So, until Smith establishes his claim (which, I am sure, he will not be able to do), we should treat blacks and whites as equals. Imagine the consequences if, instead, people claiming no relevant difference had the burden, or even if no one had the burden. All sorts of bigotry could then be easily rationalized.

The application of this to the question of whether, in the justification of moral norms and values, equal consideration ought to be given to the well-being of everyone is obvious. Unless there are relevant differences between people that justify our giving the well-being of some more consideration than the well-being of others, then, according to the principle of universalization, equal consideration is to be given to the well-being of everyone, and the burden of proof rests upon those who would claim that there are relevant differences. Since such a burden has never been met (I doubt it ever will), we can conclude therefore that, for purposes of justifying moral norms and values, everyone's well-being should indeed be considered equally. That is, equal degrees of well-being should be given equal weight, no matter whose well-being it is. This defence of equal consideration is certainly not conclusive. But, if I am right, it shows something almost as significant as a conclusive defence. It shows that we are warranted in proceeding on the premiss that equal consideration is justified until someone succeeds in showing that it is not.

Given the above two premisses—the point of morality premiss and the equal consideration premiss—we can now refute the sceptic. From these two premisses it follows that the object of morality is our well-being, with that of each of us being given equal consideration. And, as we have already seen, if Q is the object of P (e.g. if cutting is the object of knives), then the best, or most justified, Ps are those most conducive to Q. Therefore, the most justified moral norms and values are those most conducive to our well-being, giving everyone's well-being equal consideration. So, contrary to what the sceptics claim, the question, 'Which moral norms and values are most justified?', does indeed have a correct answer. The answer, once again, is: Those most conducive to our well-being, giving everyone's well-being equal consideration. Let us refer to these norms and values simply as those most conducive to 'overall' well-being.

This approach to moral, or normative, justification is not new. It is a 'utilitarian' approach, or methodology, because moral justification in terms of overall well-being is, in effect, in terms of the utilitarian standard. According to this standard, the more conducive to overall well-being an alternative is, the better, or more justified, it is. And, as always, 'overall' well-being is to be determined by giving equal degrees of well-being equal consideration, no matter whose well-being it may be. But in what sense exactly this is a utilitarian approach, or methodology, needs to be clarified in two ways.

First, utilitarians usually formulate the utilitarian standard in terms of maximizing what they call 'utility', not 'well-being'. The formulation in terms of maximizing utility has one advantage. 'Utility', being a technical concept, is (or can be made) more clear-cut than the ordinary language concept of 'well-being'. I shall introduce the concept of 'utility' later (Sect.

1.7), but a formulation of the utilitarian standard simply in terms of 'well-being' will do for now.

Second, utilitarians are of two types: direct (or act) and indirect (or rule). The distinction is crucial. Direct utilitarians use the utilitarian standard for evaluating and justifying not moral norms and values, but acts, policies, social institutions, and so on, the sort of things, in other words, that are evaluated and justified by appeal to moral norms and values. They consider the utilitarian standard (in the form of a norm) to be the fundamental norm of morality. Indirect utilitarians, on the other hand, use the utilitarian standard instead for evaluating moral norms and values. So, when it is put to this use by indirect utilitarians, the utilitarian standard cannot itself be a moral norm or value (as it is for direct utilitarians), since that which justifies somthing, M, obviously cannot itself be M, or part of M, without circularity.

Now from the two premises set out in this section we can conclude only that the utilitarian standard is to be used for evaluating and justifying moral norms and values. We cannot conclude that it is itself a moral norm or value. Thus, when I say that the methodology of moral justification that follows from these two premises is utilitarian, I mean *indirect* utilitarian, and that only. It is essential not to confuse this with the direct utilitarian approach which, as I argue in Section 1.6 below, is altogether implausible.

1.4 AN INDIRECT UTILITARIAN METHODOLOGY

The indirect utilitarian methodology suggested, but not fully spelt out, in the last section needs to be formulated very carefully; the wrong formulation and indirect utilitarianism becomes as implausible as direct utilitarianism. I shall try, in this section, to explain exactly what the correct formulation is.

We have already seen that, at the very least, indirect utilitarian methodology calls for moral norms and values to be justified, ultimately, in terms of the utilitarian standard. In other words, they are to be justified in terms of their conduciveness to well-being, with everyone's well-being to be given equal consideration. The more conducive to overall well-being a moral norm or value is, the more justified it is. So far, so good. But now we encounter a major problem: there is more than one way to interpret what it means for norms and values to be justified in terms of their conduciveness to something, P (in this case, overall well-being). The first thing that an appreciation of what I take to be correct indirect utilitarian methodology hinges upon is an understanding of the appropriate way to interpret what this means.

Perhaps the most natural and, in any case, the most common way of interpreting what it means for moral norms to be justified in terms of their conduciveness to overall well-being is that they are to be justified in terms of how conducive to overall well-being the norms would be if generally followed—that is, if successfully, and universally, complied with throughout society. But, for reasons that will become apparent (see esp. Sect. 1.5), this is definitely not the appropriate interpretation.

The appropriate interpretation of what it means for moral norms to be justified in terms of their conduciveness to overall well-being is not in terms of their conduciveness if *generally followed*, but is instead in terms of their conduciveness if *backed by social pressure* (an interpretation that will be made more precise shortly). A norm or value is 'backed by social pressure' if society exerts social pressure upon people to act according to the norm or value. This social pressure is brought about, in general, through parental training, moral education, and, more specifically, by criticizing, or even ostracizing, those who fail to act according to the norm or value, and praising those who do. And 'informal' social pressure such as this may be reinforced by 'formal' social pressure, which is the social pressure resulting from *governmentally* administered punishments and rewards.

The second thing upon which an appreciation of the indirect utilitarian methodology being formulated here hinges is an understanding of the technical concept of a 'social pressure system'. A social pressure system is an entire system of norms and values, with each being backed by social pressure of a certain intensity, form, and scope. A society's social pressure system includes all the moral norms and values that are backed by social pressure within that society, along with the amount, form, and scope of social pressure applicable to each. But, as I have explained elsewhere (Haslett 1987a: ch. 4), a society's social pressure system includes more than just its norms and values of morality. It includes as well any other norms and values that are backed by informal social pressure within that society, along with, once again, the social pressure applicable to each—norms and values such as those of etiquette and those of good taste. Thus an entire social pressure system is a complex conglomeration indeed of many, more or less interrelated norms and values, along with various degrees and forms of social pressure, with differing scopes, in support of them. But the crux of any social pressure system remains its norms and values of morality, and it is upon these that I shall focus here. Moreover, as a compromise to more familiar terminology I shall, throughout, often refer to a society's social pressure system as its 'system of morality' or simply its 'morality'.

The third and final thing upon which an appreciation of the methodology being formulated here hinges is an appreciation of the distinction between justifying a single norm or value within a system of morality and justifying the entire system. Consider, first, justifying the entire system. According to this methodology, some system of morality is more justified than any alternative system if and only if it is more conducive to overall well-being than any alternative system.

Consider, next, justifying a single norm or value within a system of morality. Justifying a single norm or value is, according to this methodology, far simpler than justifying an entire system; yet the formulation of how to do so is more complex. In the first place, the justification of a single norm or value within a system depends upon how well, as compared with its alternatives, that norm or value 'fits into' the system. And, according to this methodology, how well it fits into the system comes down in the end (roughly speaking) to how conducive to well-being the system is with that norm or value compared to how conducive it is with any alternative. Thus the justification of a moral norm or value is always—either explicitly or implicitly—a justification of it with respect to some system of morality, which we may refer to as the 'frame of reference' for its justification. (The justification of a moral norm or value is also justification with respect to a particular set of circumstances, but that goes without saying.)

Let me now be more precise about the justification for a single norm or value (as opposed to the justification of an entire system of morality). Take any single norm or value, N. Say that N is backed by social pressure of the amount, A, form, F, and scope, S. This combination shall be referred to as '$N(AFS)$'. Finally, let us take system of morality, M, as our frame of reference. According to the methodology being formulated here, $N(AFS)$ is, with respect to system of morality M, more justified than any alternative if and only if M with $N(AFS)$ is more conducive to overall well-being than M with any alternative to $N(AFS)$.

This methodology for justifying moral norms and values can be put more simply, but less precisely, as follows. Norm or value N is, with respect to system of morality M, more justified than any alternative if and only if M with social pressure in support of N is more conducive to overall well-being than M with social pressure in support of any alternative to N. If, for example, M is our own system of morality and N is, say, an exception for euthanasia that is to be built into our system's moral norm prohibiting killing, then, according to this methodology, this exception for euthanasia is more justified than any alternative if and only if our system of morality with social pressure in support of this exception is more conducive to overall well-being than our system with social pressure in support of any alternative to this exception.

Or, put even simpler, this exception for euthanasia is justified if and only if social pressure in support of it is more conducive to overall well-being than social pressure in support of any alternative. Whenever, as here, no frame of reference is explicitly mentioned, we are then to assume that the frame of reference is our society's current system of morality. (And, of course, whenever no particular set of circumstances is explicitly mentioned, we are then to assume that the circumstances are those existing here and now.) Finally, as this euthanasia example illustrates, moral justification can, for all practical purposes, focus not just on complete norms or values, but also on important *parts* of norms or values, such as the exceptions to be built into them. (For a more extensive formulation and defence of indirect utilitarian methodology, see Haslett 1987*a*: part I; alternative formulations are found in Brandt 1979; Hare 1981; Sumner 1981; Harsanyi 1982.)

1.5 AN ILLUSTRATION: JUSTIFYING THE VIRTUE OF CHARITY

Let me try to make this abstract formulation of indirect utilitarian methodology more clear by a concrete illustration of how, according to this methodology, the justification of moral norms and values is to proceed. Say we are interested in what a given system of morality should require from individuals with respect to aiding, financially, those who are starving throughout the world. Two alternatives are being considered.

The first alternative is that of including a norm within the system of morality making it obligatory to give aid up to the point of 'diminishing marginal returns'; that is, up to the point where to give any more would leave the donor worse off than those to whom the aid is directed. So, with respect to the starving, this would mean no less than giving aid up to the point where the donor was on the verge of starvation himself. Such an extremely demanding approach is essentially the one advocated by Peter Singer in his well-known article, 'Famine, Affluence and Morality' (1972). The general idea appears to be that, by giving up to the point of diminishing marginal returns, overall well-being will be maximized.

The second alternative is for charitable giving not to be required by any norm but, instead, to be in accordance with certain values—or, more specifically, virtues—ones such as 'generosity', 'compassion', and 'kindness' (recall here the distinction, explained in Sect. 1.2, between norms and values). For simplicity, let us think of the cluster of related virtues that constitute alternative II simply as one conglomerate virtue, which we may call 'charity', the opposite of which is the vice 'selfishness'. By a pattern of giving over time or, perhaps, even by a single instance of aid if spectacular enough, we acquire the virtue of charity, which entitles us to respect and admiration. By a pattern of not giving over time, on the other hand, we may acquire the vice of 'selfishness', which leaves us subject to reproach. In general, by being a virtue, charity is not an obligation; instead, in every case it is, morally speaking, optional. By being a virtue, charity is 'supererogatory', which means it is benevolence beyond the call of duty and, as such, highly praiseworthy.

So, to summarize, alternative I takes the form of a *norm* in terms of which charitable giving, up to the point of diminishing marginal returns, is always mandatory (let us refer to this norm as 'N'), while alternative II takes the form of a *virtue* in terms of which charitable giving is always optional, yet highly praiseworthy (let us refer to this virtue as 'V'). These two alternatives are not, of course, the only ones available, and a fully adequate investigation into the proper place of charitable giving in the topography of a moral system would require the consideration of other alternatives as well. One possibility, mentioned by Singer himself (1979: 180–1), is a norm that, instead of requiring that we give up to the point of diminishing marginal returns as does N, requires only that we give, each year, 10 per cent of our income. But justifying this 10-per-cent norm creates a line-drawing problem; why 10 per cent rather than, say 5 per cent, or 25 per cent? In any case, for simplicity we shall focus only on which is most justified as between N and V.

With indirect utilitarian methodology, of course, which is most justified depends (roughly speaking) upon which has the best consequences. But remember: we are not to focus on the consequences of *general compliance* with the alternatives in question. Instead, we are to focus on the consequences of there being, within the system of morality that we take as our frame of reference, *social pressure in support of them*. More precisely, we are to focus on the consequences of there being social pressure of a certain intensity, form, and scope. But to simplify matters, let us, for the present at least, assume that the intensity, form, and scope in question is that typical for norm and values of a type similar to N and V in our current system of morality. So which alternative as between N and V is most justified comes down to which, as between N and V, would have the best consequences if supported by social pressure of (we are assuming), a typical amount, form, and scope. Finally, in evaluating N and V, let us take our society's current system of morality as our frame of reference.

We might begin our investigation into which is most justified as between N and V by noticing that N—a norm making it obligatory for us to give financial aid up to the point of diminishing marginal returns—seems to cohere better than does V with the high value our morality places upon human life, any human life. On the other hand, if, before all else, we had to give up to the point of diminishing marginal returns, this would seem to leave us nothing left over for attending to the needs, beyond mere subsistence, of our own children, and perhaps even for paying our debts. Yet attending to our children's needs beyond just mere subsistence (if possible), and paying our debts, are generally viewed as obligatory. V, being a virtue in terms of which any aid is optional, not mandatory, allows us to hold money back for fulfilling these and other financial obligations. So V seems to cohere better than does N with the obligations our morality places upon us to provide well for our children, to pay our debts, and so on.

Of course if some alternative, *A*, did not cohere very well with certain norms and values, then perhaps it would be these norms or values that should be rejected (or altered), rather than *A*. That is why a frame of reference can never be more than tentative. And, according to indirect utilitarian methodology, which is preferable, as between rejecting the alternative under consideration and rejecting the norms and values with which it does not cohere, is to be determined, like everything else in the end, by appeal to the utilitarian standard.

Let us now assume that, after an appropriate investigation, we conclude that both alternatives, *N* and *V*, cohere equally well. The next step is to investigate the consequences of each alternative in terms other than merely its coherence. The next step, in other words, is to investigate the extent to which each alternative would be conducive to our well-being in ways other than merely by 'fitting in' well with our morality. And, once again, our focus is to be not upon the consequences of general compliance with the alternative, but upon the consequences of there being social pressure in support of them.

Actually, the line between these two 'steps' would rarely, if ever, be clear-cut; the investigation below into which alternative would have the 'better consequences' could perhaps be viewed as a continuation of the investigation into coherence, or the investigation into coherence could perhaps be viewed as an investigation into which would have the better consequences, thus collapsing these steps. But there is never, in practice, any need to distinguish clearly between these steps anyway.

I suggest that social pressure in support of the second alternative, *V*, would have the better consequences. Let me explain why, although, once again, in no more detail than necessary for illustrating indirect utilitarian methodology. The first alternative, the norm, *N*, that requires us to contribute up to the point of diminishing marginal returns, requires too much. As long as people are starving who, through our contributions, we can save, this norm requires that we contribute all that we earn beyond merely what we need for our own basic necessities. This would mean nothing left over for automobiles, television sets, fancy clothing, nights out 'wining and dining', trips to the beach, and so on. I seriously doubt if people, generally, could ever be motivated, by social pressure of 'typical' intensity, to give anywhere near as much as this. People instead would simply disregard this norm, making a mockery of it, which not only would be unfortunate for the starving, but might tend even to lessen people's respect for morality in general. And sooner or later, of course, the norm would simply drop out of society's morality altogether, thus leaving those who are starving with perhaps no normative support at all.

It might, of course, be possible to motivate people to comply with N if it were backed with social pressure of not just typical, but *extraordinary* intensity—social pressure achievable only through some Draconian, state-administered punishment for non-compliance, such as, say, death by being boiled alive. But social pressure this intense would only make matters worse. By forcing people to give away all that they earned beyond merely what they needed for their basic necessities, once they had earned enough for their basic necessities people might well simply stop working. This would be especially true of people whose jobs were dull and tiring, as most people's jobs are. 'Why', these people might ask, 'should I go on working hard if I can't keep what I earn.' Perhaps these people could be coerced into continuing to work, but that would amount to slavery—hardly a satisfactory solution. And, in any case, people could not very well be coerced into working creatively or using their talents in a way that was socially optimal. In some hypothetical, ideal world people might give as much as N required while continuing to work just as hard and creatively as ever but, in the real world, most people simply do not have, and could not be made to have, the will-power for this. Most people are not saints. And all people are fallible. So the result of this extraordinarily intense social pressure would probably be not that the starving nations of the world would eventually approach the wealth of the rich nations, but that (due to a loss in people's motivation to work) the rich nations would eventually approach the poverty of the starving nations.

And any social pressure this intense obviously would have other costs as well. These would include costs in the form of whatever energy and expenses were necessary for administering the severe punishment in question, costs in the form of whatever suffering those experiencing this punishment had to endure, and costs in the form of whatever anxiety the threat of this punishment created throughout society.

Consider now the second alternative, V. With this alternative giving is not required by a norm, but is encouraged by its being a virtue; so the emphasis is not upon punishments for failing to give, but upon rewards, in the form of praise and respect, for giving. The costs of effective social pressure would therefore be far less; no expensive, Draconian punishments would be necessary; the emphasis upon rewards would cause little suffering, anxiety, and so on. Most importantly, since, with V, giving is not mandatory, but optional, this approach would not unrealistically strain human capacities; it is, in other words, more compatible with human fallibility. People, knowing they would not be forced to give away anything they did not freely choose to give away, would continue working hard and creatively. So more would be produced. With more being produced, more would be available to give to those in starving nations, and no doubt would, in fact, be given. Thus, V would very likely end up being more beneficial than N not only to those in rich nations, but, somewhat ironically, to those in starving nations as well. So I think that social pressure in support of V would indeed have better consequences. If I am right, then V is, in terms of the methodology defended here, more justified.

The example of N versus V illustrates the advantages of an indirect utilitarian methodology, such as the one defended here, that focuses on the consequences of social pressure rather than on the consequences of general compliance. The consequences of social pressure in support of a norm include not only the consequences of any resulting successful compliance, but also the consequences of any resulting *failures* to comply. Recall here the disastrous consequences of widespread failures to comply with N. Focusing on the consequences of social pressure thus allows us to take any such human fallibility fully into account. And this focus allows us, as well, to take fully into account the costs of social pressure (in time, money, energy, and suffering). Contrast this with focusing instead on the consequences of general compliance. By focusing on the consequences of general compliance—that is, on the consequences of universal success in complying—the consequences of human fallibility and the costs of whatever social pressure might be needed for minimizing these consequences, will fail to be taken into account. Look at it this way. No norm will ever come even close to being complied with universally; there will always be those who fail to comply. Thus the consequences of general—that is, universal—compliance will never occur; these consequences (as opposed to the consequences of social pressure) are altogether moot. Since these consequences are altogether moot, focusing on them thus has no point.

Another advantage of focusing on the consequences of social pressure rather than of general compliance is that we are not thereby limited only to comparing alternative norms; we can, as we did with N and V, also compare a norm with a value (virtue or ideal). Were we focusing instead upon general compliance, we could not compare N with an alternative in the form of a value, such as charity, since, strictly speaking, values are not, like norms, the sort of criteria that prescribe, and thus may legitimately be said to be 'complied with' (recall again the distinction between norms and values explained in Sect. 1.2). Yet values and norms are alike in being backed by social pressure; thus a comparison in terms of the consequences of being backed by social pressure can encompass both.

1.6 INDIRECT UTILITARIANISM VS. DIRECT

In order to complete the case for this indirect utilitarian methodology, the standard objections against indirect utilitarianism must now be examined.

Curiously, the most common objections against indirect utilitarianism are that it fails to overcome the objections against *direct* utilitarianism. Thus, to understand and evaluate objections against indirect utilitarianism, we must first see what the corresponding objections against direct utilitarianism are.

As will be recalled, what distinguishes direct (or act) from indirect (or rule) utilitarians is the use to which they put the utilitarian standard. Indirect utilitarians put it to use not as a moral norm or value. Instead, they put this standard to use for evaluating, and thus for justifying, moral norms and values. Direct utilitarians, on the other hand, put the utilitarian standard (or its analogue in the form of a norm) to use directly as a moral norm itself—indeed, as the fundamental moral norm. Take, for example, deciding whether or not to steal Jones's car. The indirect utilitarian will use the utilitarian standard for evaluating alternative moral norms dealing with theft and will advocate compliance with the alternative that is evaluated most highly, which, I think, will turn out to be a norm something like: 'Do not take what does not belong to you, except with the owner's consent or in an emergency', a norm which prescribes that we not steal the car. For the direct utilitarian, on the other hand, the utilitarian standard itself takes on the role of that moral norm which we are to use, directly, for deciding whether to steal Jones's car, and this standard (or its analogue in the form of a norm) prescribes, of course, that we steal the car if and only if doing so maximizes overall well-being or utility. Thus, according to the direct utilitarian, perhaps we should steal Jones's car, depending solely upon whether doing so maximizes utility or not. In general, we should, according to the direct utilitarian, choose any act, policy, or social institution if and only if it maximizes utility.

This, then, gives rise to the most popular objection against direct utilitarianism. Direct utilitarianism, so this objection goes, prescribes that we kill, lie, punish the innocent, enslave minorities, or do who knows what other radically counter-intuitive things, provided only that they maximize utility. And by 'counter-intuitive' things, what is meant is simply things contrary to our moral feelings or 'intuitions'. Let us call this the 'popular' objection.

Contrary to most philosophers, however, I do not find the popular objection to be conclusive. For all we know, the only circumstances that would (in direct utilitarian terms) ever justify killing, lying, punishing innocent people, enslaving minorities, and so on, might be so dire that either these circumstances would never occur or, if they ever did, the case for doing these things in circumstances that dire would be so compelling that doing them would no longer seem counter-intuitive. (For a development of this point, see Hare 1981: ch. 8.) In other words, what maximizes utility in the circumstances at hand is always so difficult to determine that, for all we know, what direct utilitarianism actually does prescribe might never turn out to be radically counter-intuitive after all. Besides, there is no reason to assume that our moral intuitions are always reliable anyway. So the popular objection is hardly conclusive. But the difficulty of determining what maximizes utility gives rise to an objection that is conclusive. Let me explain.

Direct utilitarian calculations—calculating which alternative maximizes utility in the circumstances at hand—are indeed difficult and thus unreliable. This is a consequence of (a) the severe time constraints under which such calculations must often be made, (b) the necessarily limited amount of information upon which they are usually based, and (c) the rationalizations to which they are subject. In general, the unreliability of utilitarian calculations is a consequence of human fallibility. Because of the difficulty and thus unreliability of utilitarian calculations, we would often be unable to anticipate how other people's calculations were going to turn out, and thus what act of theirs they were going to end up believing to be the one that maximized utility. Take, for example, someone, Jones, who sympathizes with those who are starving throughout the world, and who dislikes you. If you were to loan Jones money, would he consider it utility maximizing to pay you back or to give the money instead to the starving? Or, if you were in the woods near a cliff alone with Jones, would he consider it utility maximizing to rid the world of someone he thinks undesirable by pushing you off? And, if you were to ask him these questions, would he consider it utility maximizing to tell you the truth? How could you ever really know? In short, if everyone tried all the time to do only that which maximized utility, as direct utilitarianism prescribes, there would be little uniformity in behaviour from one person to the next with respect to morality, and thus little basis for ever predicting how others were going to behave towards us. This would undermine the very purpose for which we have a morality in the first place, which is to make us better off through being more secure and having a basis for trusting other people.

No objection against direct utilitarianism could be more conclusive than this. Like all utilitarians, direct utilitarians have, as their goal, that overall well-being or 'utility' be maximized. Yet, since direct utilitarianism undermines the very purpose of morality, with direct utilitarianism—which calls for everyone, always, to maximize utility—utility is not maximized. In short: direct utilitarianism is self-defeating. It is often said, and with some justification, that one cannot very well realize true happiness for oneself by pursuing it directly; rather, one realizes it by pursuing, instead, other goals, such as the happiness of others. This is called the 'paradox of hedonism'. Similarly, the fact that overall utility cannot very well be realized by everyone's pursuing overall utility directly might well be called the 'paradox of utilitarianism'.

Having thus examined the main objections against direct utilitarianism, we can now appreciate the standard objections against indirect utilitarianism. One of the most commonly repeated objections stems from David Lyons (1977). It is that indirect utilitarianism will end up justifying moral norms that prescribe the very same acts as does direct utilitarianism. In other words, indirect utilitarianism 'collapses' into direct utilitarianism; the two turn out to be 'extensionally' equivalent (to use a technical term). And if they are extensionally equivalent, it follows that indirect utilitarianism must be as counter-intuitive, and thus as subject to the so-called popular objection, as direct utilitarianism.

The argument for showing that indirect (rule) utilitarianism collapses into direct (act) utilitarianism is probably sound for an indirect utilitarian methodology that justifies moral norms in terms of the consequences of general—i.e. universal—*compliance* with the norms. As we have seen, however, the methodology proposed here instead justifies moral norms in terms of the consequences of their being backed by social pressure. Thus, rather than rehearse the argument for showing why an indirect utilitarianism that focuses on the consequences of universal compliance collapses into act utilitarianism, let me show instead why an indirect utilitarianism that focuses on the consequences of being backed by social pressure does *not* collapse into direct utilitarianism.

As we have seen, an indirect utilitarian methodology that justifies moral norms in terms of the consequences of their being backed by social pressure takes human fallibility fully into account. Since, by taking human fallibility into account, we found that the consequences of the direct utilitarian system of morality would be disastrous, it follows that the direct utilitarian system is not justifiable in terms of indirect utilitarian methodology. What system of morality then is justifiable in terms of this methodology? It must, we know, be a system that is suitable for beings who are fallible; that is, for human beings. This means that the system must contain only a limited number of moral norms and values (fallible beings cannot handle an unlimited number). And, for the most part, the norms must meet two conditions of simplicity: that of being easy to learn and that of being easy to comply with. For moral norms to be easy to learn, they must be ones that are relatively simple to state, with only a relatively few exceptions built into them. And for moral norms to be easy to comply with, they must be ones whose prescriptions people, generally, do not find very difficult to figure out or be motivated to perform. The only sort of moral norms that can meet these conditions of simplicity and still be compatible with the point of morality in every other respect are, I submit, ones that, for the most part, are rather similar to just the sort of moral norms most of us have been brought up to observe, moral norms requiring that we tell the truth, pay our debts, refrain from violence, theft, and so on. That is to say, they are just the sort of moral norms that tend to agree rather well with most people's moral intuitions. But simple, easy-to-comply-with moral norms such as these obviously do not always prescribe the very same acts as does the difficult-to-comply-with utilitarian criterion; in other words, these norms obviously do not always prescribe the same acts as does direct utilitarianism. This then is precisely why the version of indirect utilitarianism being proposed here does not 'collapse' into direct utilitarianism.

Contrary to what some critics of indirect (or rule) utilitarianism seem unable to understand (e.g. Smart 1973), adherence to these simple, easy-to-comply-with norms even when one believes that contravening them would maximize utility is not mere 'rule worship'. The point, once again, is this: because of mistaken calculations and rationalizations, because of the unpredictability in people's behaviour and thus insecurity throughout society that would inevitably result, in short, because of human fallibility, for people to take it upon themselves to contravene these norms whenever they believed that doing so maximized utility would cause far more harm than good. (On this, see Haslett 1987a; part II.)

I have been arguing that the moral norms justifiable in terms of this indirect utilitarian methodology are ones that tend to agree rather well with most people's moral intuitions. If I am right, then it follows that the popular objection against direct utilitarianism—that it prescribes acts radically contrary to most people's moral intuitions—is not applicable to this version of indirect utilitarianism (or, more precisely, to the moral norms justifiable in terms of it). Some critics of indirect utilitarianism, however, do not agree. They argue that, although the moral norms justified in terms of indirect utilitarian methodology might indeed come closer than direct utilitarianism to prescribing acts that agreed with our moral intuitions, they nevertheless would not come close enough to avoid the popular objection. This methodology, these critics say, would justify moral norms, and rights, that were seriously skewed against the interests of those in certain groups—norms that, even though they might not prescribe the very same acts as direct utilitarianism, would nevertheless be quite contrary to our intuitive understanding of justice.

For example, Allen Buchanan argues that any moral right to a decent minimum of health care justified in terms of indirect utilitarian methodology would not be applicable to all alike (as, intuitively, we think it should), but would exclude certain groups from coverage, such as those with Down's syndrome (1985). The reason those with Down's syndrome would allegedly be excluded is that they, in order to have a decent minimum of health care, typically require 'a large expenditure of social resources over a lifetime' while, at the same time, they, being mentally retarded, typically make less of a contribution to social resources than most do. Therefore, Buchanan concludes, so as to conserve as much of society's scarce resources as possible, an indirect utilitarian right to health care would exclude those with Down's syndrome.

This conclusion is mistaken. First of all, Buchanan relies upon a cost-benefit analysis in terms of 'social resources' (i.e. dollars). Let us not, however, be misled into thinking that, according to indirect utilitarian methodology, a cost-benefit analysis in terms of social resources is the kind of analysis that, ultimately, counts. What ultimately counts is a cost-benefit analysis in terms of well-being or utility. Once we look beyond mere dollars, or social resources, to well-being itself, it will no longer appear so obvious that indirect utilitarianism calls for the exclusion of those with Down's syndrome. In fact, a proper indirect utilitarian analysis, one in terms of well-being itself, reveals, I submit, that the exact opposite is true; that any such right includes everyone, without regard to Down's syndrome or anything else. Although including those with Down's syndrome might require that taxpayers each pay a few dollars more in taxes per year, the loss of well-being that these few dollars represented for any given taxpayer would be so minuscule that even the combined loss in well-being that taxpayers suffered would be small indeed compared to the combined gain in well-being that those with Down's syndrome enjoyed from being given minimum care like everyone else—not to mention the gain to their families and loved ones. (On this, see also Brandt 1979: 316–19.)

But what is even more important is this. Human rights that blatantly exclude some such as those with Down's syndrome set an example throughout society of cruelty, bias, insensitivity to the suffering of others, and rejection. Conversely, the example that not excluding anyone sets is that of compassion, impartiality, sensitivity, and a certain 'togetherness'. The latter are values that must be taken seriously throughout society for the sake of *everyone's* well-being. We must never forget the constraints of coherence. Human rights must cohere well with these essential values; must reinforce them. Any society that, just to save a little money, disregards the need to thus reinforce these essential values makes a grave mistake, one that can never, as Buchanan claims, be justified in terms of indirect utilitarian methodology. So I conclude that the moral norms justifiable in terms of this methodology are not counter-intuitively skewed against the interests of those in certain groups, as some critics claim, and, therefore, indirect utilitarianism does in fact avoid the popular objection against direct utilitarianism.

But, it might be asked finally, does indirect utilitarianism really overcome what I am taking to be the *conclusive* objection against direct utilitarianism? This, it will be recalled, is the objection that, because of the difficult calculations called for by direct utilitarianism, predictability and uniformity in people's moral behaviour would be impossible. After all, it might be argued, indirect utilitarianism calls for calculations that are equally difficult; the only difference being that, with direct utilitarianism, the calculations concern the consequences of doing particular acts, whereas, with indirect utilitarianism, they concern the consequences of social pressure in support of particular moral norms and values. With direct utilitarianism, you would not know whether someone, Jones, believed the particular acts of reneging on a debt to you, pushing you off a cliff, and lying to you maximized overall well-being, whereas, with indirect utilitarianism, you would not know whether Jones believed that social pressure in support of moral norms allowing these very same acts maximized overall well-being. Thus, it might be concluded, indirect utilitarianism entails a lack of predictability and of uniformity in people's moral behaviour no less than does direct utilitarianism.

My reply to this argument is, in short, that it overlooks the role of social pressure in people's moral behaviour. People will, no doubt, have all sorts of strange and unpredictable beliefs about what moral norms are most justified, just as many people have strange and unpredictable beliefs about what laws—for example what tax laws—are most justified. In the case of law, formal social pressure, in the form of state-administered punishment for non-compliance, normally generates sufficient compliance in spite of these differing beliefs. Likewise, informal social pressure, reinforced, in some cases, by formal social pressure, normally generates sufficient compliance with moral norms in spite of many differing beliefs. Because of this social pressure, a person's atypical beliefs about which moral norms are most justified, even when quite contrary to majority beliefs, often do not affect, to a great extent, which moral norms that person complies with. A person's atypical beliefs normally affect, instead, which moral norms that person helps support by means other than personal compliance, by means such as rational argumentation, and criticism or praise of other people's behaviour. This is because, generally speaking, people are under far less social pressure with respect to which norms they are to *help support* than with respect to which norms they are to *comply with*—and this is as it should be in a society dedicated to the free expression of beliefs. Compare, again, how it is with law: people may, with impunity, argue as much as they please against current tax law and take actions in support of changing it, provided only that, in the mean time, they continue to pay their taxes. If enough people come to believe a current moral norm is unjustified, social pressure will, of course, eventually shift in favour of an alternative. Therefore changes in what norms and values are backed by social pressure do, fortunately, occur—fortunately because, otherwise, there could be no moral progress. But changes occur gradually enough so as not to result in a significant lack of predictability or of uniformity in people's moral behaviour. So, in sum, once we understand the role social pressure plays in people's moral behaviour, we see that the second objection to direct utilitarianism mentioned above is not applicable to indirect utilitarianism either. (Cf. Haslett 1987*a*: sects. 4.4 and 10.5, and ch. 5.)

1.7 WELL-BEING AND UTILITY

So far I have tried to show that, by way of indirect utilitarian methodology, questions about which moral norms and values are most justified have correct answers. But, it might now be objected, in claiming that this methodology yields correct answers, I am overlooking one huge problem. This methodology leans heavily upon the concept of 'well-being' and if we have no adequate understanding of this concept, if no analysis of well-being can be shown to be any more justified than any other, then it would appear that there are no correct answers after all. So to complete the argument against the sceptic, to substantiate the claim that indirect utilitarian methodology does, in principle, yield correct answers, I must try now to provide a satisfactory analysis of the concept of 'well-being'.

Throughout the history of philosophy, many different sorts of things—self-realization, autonomy, health, and so on—have been put forth as what constitutes human well-being or welfare. I cannot here, of course, investigate all of these different approaches. I shall instead concentrate only on the approach that is today most widely accepted among economists and other social scientists, the approach that happens also, in my view, to be the right one. This is the approach that analyses personal well-being in terms of what is called personal 'utility'.

It is important to understand, however, that those who take this approach do not deny that these other things—self-realization, autonomy, health, and so on—are also important constituents of well-being. They deny only that these other things delineate well-being in the most abstract—and fundamental—sense possible.

But, among those who analyse well-being in terms of utility, there are competing views about what, exactly, 'utility' is. In what follows, I shall examine the main competing views, or 'models' of utility and try to determine which is correct. By the 'correct' model of utility, I mean simply the one that does, in fact, help clarify our ordinary, everyday concept of well-being.

A terminological note is necessary before proceeding. I shall, throughout, be using the expressions 'personal well-being', 'personal welfare', and 'personal interests' interchangeably. There are indeed subtle distinctions in ordinary language between these three, but I do not find it useful to distinguish between them here. Those who do may choose their favourite from among the three, and substitute it for the other two throughout. My only request is that they choose one that, for them, designates a person's *self*-interest, rather than one that, for them, serves as a general term designating anything the person may value, view as 'good', or be interested in. This is crucial. For most people at least, what they value, view as good, or are interested in includes what is, or what they think is, in their own self-interest, but is not limited to that. I shall be using the expressions 'personal well-being', 'personal welfare', and 'personal interests' interchangeably to designate a person's *self*-interest, and that only.

One more terminological note. Mainstream utilitarians typically distinguish between *personal* utility and *social* utility. I shall do likewise. The general idea is that personal utility has to do with personal well-being, and social utility is the sum total, or arithmetic mean, of personal utility throughout society (e.g. see Harsanyi 1982: 40; Sumner 1981: 178).

As has been frequently pointed out (Glover 1977; Grice 1967; Griffin 1986; Sumner 1981), most philosophers and social scientists adopt some version of either one or the other of two standard models of personal utility: the experience model and the preference model. Let us examine each of these in turn.

1.7.1 The Experience Model

According to all versions of the experience model, personal utility is defined in terms of having certain experiences—that is, certain mental states or states of consciousness. The traditional version of this model identified these experiences with pleasure or happiness (see e.g. Bentham 1789).

Before long, however, this version of the experience model was found to be objectionable. Among the objections to it are that it restricts what may be said to constitute personal utility too narrowly, since pleasure and happiness are not the only experiences that may be said to be of value. For example, the experience one has in contemplating a beautiful sunset, or in doing something worthwhile for others, may not, strictly speaking, be accurately describable as 'pleasure', or even 'happiness', yet these experiences are nevertheless of value. And, as John Stuart Mill said, it may be better to be Socrates dissatisfied (i.e. unhappy) than to be a pig satisfied (i.e. happy). Moreover, it is unclear exactly what is meant by happiness, and even whether it can be appropriately classified as an experience(s). Other critics, especially social scientists, have objected to the impossibility of measuring pleasure and happiness precisely. By a precise measurement, they mean a so-called cardinal measurement, one that allows units of pleasure or happiness to be added together meaningfully, so that the mathematical calculations necessary for determining which alternatives maximize utility can be performed. Related to the problem of how, precisely, to measure the pleasure or happiness of any one person is the problem of how, validly, to compare the pleasure or happiness of one person with that of another (i.e. the problem of 'valid interpersonal comparisons'). These problems remain unsolved.

And there is still another problem with the experience model, a more subtle one. Say that Mary loses her entire fortune in the great Black Monday stock-market crash, but, being at the time on a safari in Africa, she does not yet know what has happened. If personal well-being equalled utility, and utility were defined in terms of having certain experiences, then she could not have been made worse off by the crash before it affected any of her experiences. Yet clearly she was made worse off by the crash before it affected any of her experiences; thus, the traditional experience model seems deficient in this respect as well.

1.7.2 The Preference Model

In an attempt to avoid these alleged problems with the traditional experience model, many philosophers, and most social scientists, have now adopted instead the preference model of utility. According to the preference model, utility is preference satisfaction; the greater the number, and the strength, of a person's preferences that are satisfied, the greater that person's utility. And since, according to most versions of this model, personal utility is equated with personal well-being, the greater the number, and the strength, of a person's preferences that are satisfied, the better off that person is.

Some versions of this model define utility instead in terms of desire satisfaction. But since the person who *prefers* A to B would, I presume, *desire* A more than B, I suspect that these two variations of this model come down to much the same thing. Finally, some versions of this model add what might be referred to as a 'fully informed' requirement, which is that, in order for the satisfactions of a person's preferences to count as utility, these preferences must be the ones that the person would have if fully informed of all relevant facts. We may refer to those versions of the preference model that do not include a fully informed requirement as 'strong' versions, and those that do as 'weak' versions.

The main idea behind all versions of the preference model is simple: what increases or decreases one's utility, and thus is in one's well-being, is a matter solely of what one *oneself* prefers, not what someone else might think is best for one. This perhaps helps explain the model's popularity. It seems to reinforce that belief in toleration central to modern liberalism; that is, by making everyone's *own* preferences the final criteria of what is best for him or her, this model suggests that everyone's lifestyle should therefore be tolerated as long as it does not violate the rights of anyone else.

Moreover, this model seems to avoid most of the major objections to that version of the experience model which identifies utility, and personal well-being, with pleasure or happiness. As we saw, one of these objections is that, by limiting the experiences that are in a person's interests to just pleasure or happiness, this model restricts what is in a person's interest too narrowly. The preference model, on the other hand, can hardly be accused of restricting personal well-being too narrowly; many versions of this model place no limitations at all upon what preferences the satisfaction of which supposedly increase a person's utility and thus are in the person's interests. Another, more technical, objection to experience models has been that pleasure and happiness do not admit of a cardinal measurement. Preferences do not either, but an *ordinal* measurement of the person's preferences is possible—that is, a measurement that ranks these prefer-ences. And for determining which alternative would maximize a person's preference satisfaction, an ordinal measurement appears to be all that is needed. In short, the preference model seems to get around some of the problems of measurement and, thereby, to fit in better with much of the mathematics that economists and other social scientists want to use.

And, last but not least, the preference model seems to do better than the experience model with cases such as those of Mary, who lost her entire fortune in the stock-market crash on Black Monday while she was on a safari in Africa. It is perfectly natural to say she was made worse off the moment she lost her fortune even though, at that particular moment, her experiences had not yet been affected in any way. Advocates of the experience model, as we saw, have difficulty explaining this. Advocates of the preference model, on the other hand, can explain it easily. Because Mary, we may assume, strongly preferred that her fortune remain intact, the moment her fortune was lost this preference was frustrated, and hence she was made worse off whether her experiences had been affected or not.

Largely for these reasons, a definition of utility, and thus personal well-being, in terms of preference satisfaction is today standard among social scientists. As one textbook in economics puts it: 'What modern economists call utility reflects nothing more than rank ordering of preference' (Hirshleifer 1976: 85; see also Pearce 1986; 437; but for some reservations, see Robinson 1962; Sen 1977). And, although nothing is standard among philosophers, the preference model is more popular in philosophy these days than any other model. Philosophers who have adopted some variation of this model include Ronald Dworkin (1978), David Gauthier (1986), James Griffin (1986), Jan Narveson (1967), and Rolf Sartorius (1975).

1.7.3 The Compromise Model

Let me now introduce one more model for consideration. This model amounts to a kind of compromise between the traditional experience model and the preference model in that it draws from each but, as I shall argue in Section 1.8, is superior to both. Although what appear to be variations of this compromise model have been proposed before (Hare 1981; Haslett 1987*a*; Lewis 1946; Mill 1861; Sidgwick 1907; Smart 1973; Sumner 1981), social scientists, as well as philosophers, are not yet, I think, fully aware of the considerable advantages of this compromise.

From the experience model I take the idea that personal utility is to be analysed in terms of personal experiences. Strictly speaking, therefore, the compromise model I am proposing qualifies as a version of the experience model. (I trust that my using the word 'experiences' throughout in the most general sense possible—a sense in which it refers to any sentient state whatever—will cause no confusion.)

Contrary to those versions of the experience model considered earlier, however, this compromise model does not identify utility with any particular kind of personal experience, such as pleasure. Rather, it identifies a person's utility with whatever personal experiences that person would prefer. This ranking in terms of preferences is what I take from the preference model.

But, contrary to all versions of the preference model, it is a person's preferences for experiences, and experiences only, that, with the compromise model, delineate the person's utility. Moreover (and this is important), I am referring here to a person's preferences as between *particular* experiences, not *generic* experiences. The distinction between, on the one hand, a particular experience (e.g. the experience of having swam in the pool in one's backyard from 3:00 to 3:30 p.m., 1 August 1990) and, on the other hand, a generic experience which the particular experience happens to exemplify (e.g. the experience of going swimming) may be referred to, technically, as the distinction between an experience 'token' (a particular experience) and an experience 'type' (a generic experience). The compromise model focuses upon a person's preferences as between particular experiences only; preferences as between generic experience have no relevance whatever. Let us call this focus upon preferences as between particular experiences, and these only, the 'experience' requirement.

And, contrary to the strong version of the preference model, it is not necessarily a person's preferences as they *are* that delineate the person's utility; it is the person's preferences as they *would be* if these preferences were fully informed. Let us call this the 'fully informed' requirement. By a fully informed preference for any one, particular experience *A* over any alternative, particular experience *B*, I mean a preference that (1) is influenced by perfect knowledge of both *A* and *B*, and (2) is influenced by nothing else. This knowledge must be as perfect as it would be if the person possessing it actually had had experiences before that were qualitatively identical to *A* and *B* and had retained perfect memory of them. Let us call this the 'perfect-knowledge' requirement. And—so as to exclude irrelevant influences, such as irrational impulses, superstitions, prejudices, and the like—for a preference to count as fully informed, it cannot have been influenced by anything other than this perfect knowledge. Let us call this the 'exclusory' requirement. Thus a preference meets the fully informed requirement if and only if it meets both the perfect-knowledge and the exclusory requirements.

To try to avoid any misunderstanding, perhaps I should pursue a little further what is meant here by excluding any influences upon our preferences for experiences other than perfect knowledge of the experiences themselves. This exclusory requirement not only rules out any irrational impulses, superstitions, and the like, from influencing our preferences for experiences, but rules out as well our personal history, conceptual apparatus, and personal tastes from influencing these preferences. It is true that the subjects of these preferences—the experiences themselves—must reflect the influence of personal history, conceptual apparatus, and tastes (that is, the personal history, conceptual apparatus, and tastes of whomever's experiences they are in reality). 'Raw' experiences uninfluenced by these things are impossible. But rather than being designed to rule out the influence of these things upon the experiences themselves, what the exclusory requirement is designed to rule out is the influence of these things upon our *preferences* for the experiences. Of course these things, in so far as they do help shape the very nature of the experiences themselves, will thus necessarily influence our preferences indirectly, but the exclusory requirement rules out any direct influence.

Let me try to clarify this through an example. Say we know that a particular pleasure was caused by some perverse act, yet the person experiencing the pleasure happens to be a rather calloused individual, and therefore the perversity of the cause, in this case, played absolutely no role in shaping the nature of the pleasure itself. And say that we also know that a particular pain was caused by some heroic act, but that, once again, the cause played no role in shaping the nature of the experience itself. Now having the values we have, we may well, in real life, allow their causes to influence our overall preference between the two experiences, so that we end up preferring the pain even though, had our preference been influenced only by the nature of the experiences themselves, we would instead (by hypothesis) have preferred the pleasure. It is precisely this kind of direct influence upon our preferences that the exclusory requirement is designed to rule out.

The exclusory requirement presupposes, of course, that our preferences for experiences can indeed be based upon nothing more than the 'brute desirability' of the experiences themselves. But, in principle, they can be, even if rarely, if ever, in practice. Take, for example, the experiences of some given person, *P*. Consider, in particular, *P*'s experience of taking a warm bath early yesterday and his experience of suffering from an excruciatingly painful toothache later in the day (and *P*, incidentally, is not a masochist). Clearly these experiences need not have been *our* experiences for us to have a preference between them. And, for us to have a preference between them, it is not 'logically' necessary that we resort to any of our own background history, conceptual apparatus, or personal tastes. A basis for a preference between them is inherent in the very nature of these particular experiences *themselves*: quite simply, one is pleasurable, the other painful, and that is all there need be to it. In fact, given that his preference was fully informed, even the masochist would necessarily have to prefer the pleasure of the bath to the pain of the toothache for, if his preference was fully informed, then it would be based on, and only on, this pleasure and this pain exactly as they had been experienced by *P* yesterday, who, as I said, is not himself a masochist. And so, it goes, I am claiming, for any particular (as opposed to generic) experiences: a basis for preferences with respect to them is inherent in the very experiences themselves.

The meaningfulness of us ever being able to have a preference between what would be the experiences of someone else, *P*, might be objected to on the grounds that philosophers have not yet succeeded in showing how our having any knowledge of other people's minds, and thus of other people's experiences, is possible. But this objection would be foolish; it would be to confuse our inability to solve the philosophical problem of how we do *X* with our inability to do *X* itself. Clearly we are able to know something about other people's experiences, whether philosophers have or have not succeeded in showing exactly how. Perhaps our knowledge can never be exact, but our *approximate* knowledge about other people's experiences is, in most cases, quite sufficient for intelligent estimates about how these experiences compare, estimates close enough for most practical purposes.

To complete my formulation of the compromise model, one final matter must still be addressed: how exactly, according to this model, is personal utility related to personal well-being? The relationship between them is that what is in one's personal welfare—or, in other words, conducive to one's well-being—is not *identical* with utility (that is, with one's having certain experiences); rather, what is in one's personal welfare is what *results* in utility (that is, results in one's having certain experiences). And the experiences in question here are, of course, those that would satisfy one's fully informed preferences for experiences. So, I am claiming, what is in one's personal welfare comes down to this: something (an event, act, object, or state of affairs) is in one's personal welfare to the extent, and only to the extent, that it results in one's utility (or prevents one's disutility). And, of course, to be in one's personal welfare, an event, act, object, or state of affairs need not necessarily result in one's utility directly, but may result instead in still another event, act, object, or state of affairs that, in turn, results in one's utility. All that is necessary, in other words, is that it result in one's utility (or prevent one's disutility) ultimately. Painfully studying for a test, for example, may not result in any utility for one directly, but remains in one's personal welfare to the extent that it results in utility for one ultimately.

And if something is in one's personal welfare to the extent, and only to the extent, that it results in one's utility (or prevents one's disutility), then it is *likely* to be in one's personal welfare to the extent, and only to the extent, that it is *likely* to result in one's utility (or prevent one's disutility). Or, rather than speaking about what is likely to be in one's welfare, we can instead speak about what, relative to the information currently available, is in one's welfare. Either way we are, according to this model, referring to what is likely to result in one's utility (or prevent one's disutility). Moreover, what we mean by 'likely to result in one's utility (or prevent one's disutility)' can, with this model, be made more precise by substituting for it the technical concept of 'expected utility'. With this substitution, something is then 'likely' to be in one's personal welfare to the extent that it has expected utility for one. And the expected utility of something is found by multiplying the utility (or disutility) of each of its possible outcomes times the probability of that outcome occurring, and adding these products together. Expected utility thus represents the most favourable combination of probabilities and outcomes. Of course it may be impractical, perhaps even impossible, to represent the utility of an experience numerically—that is, with a numeral on which these mathematical operations can be performed—but this is in no way a barrier to calculating expected utility. As is explained elsewhere (Haslett 1987*a*: sect. 2.2; Lewis 1946: ch. 17, sect. 13), if utility is defined in terms of experiences, then all that is really needed is a so-called 'ordinal', as opposed to 'cardinal', measurement. These, however, are details that need not concern us here. (If utility is equated with anything less fundamental than the satisfaction of fully informed preferences for experiences—if, for example, it were to be equated with money—then certain strategies other than that of maximizing expected 'utility' might well be appropriate, but this need not concern us here either.)

Once we realize that the relationship between personal well-being and personal utility is not to be taken as one of identity, but as one of cause and effect, then the Black Monday example is no longer troublesome. Mary was made worse off by the stock-market crash because of the negative experiences (or the absence of positive experiences) that were likely for her as a result of the crash. In other words, once the proper relationship between personal well-being and personal utility is understood, we see that it is perfectly compatible with an experience model of utility for one's personal well-being to decrease or increase without there being any immediate effect upon one's personal experiences (i.e. upon one's utility).

One final point. If, as the compromise model tells us, utility is best defined in terms of preferences for experiences, then why, it might be asked, does so little everyday talk about preferences focus on preferences for experiences? Take, for example, an Olympic runner: she might talk of preferring to win her race rather than lose it; she would not be likely, instead, to talk of preferring the *experience* of winning it to that of losing it. Why? Why, in other words, is her preference talk, as well as her personal well-being talk, likely to focus on the *events* of winning and losing, not the experiences of winning and losing? At least part of the answer, I think, is as follows. Winning the race—that is, the event of winning—has far more ramifications for her than does the mere experience of winning. By winning the race she might gain, in addition to the experience of winning, a lucrative endorsement contract, the admiration of the public, more attention from the person she loves, and so on. And all of these additional gains would be likely, in turn, to result in 'satisfying' experiences that are of far greater utility for her than the mere experience of winning—gains that it would make sense to say 'resulted' from the *event* of winning, but that could hardly be said to have 'resulted' from the *experience* of winning. In other words, the event of winning is likely to be far more important overall for her well-being than the experience of winning, since the event of winning not only results in the experience of winning, but (unlike the experience of winning) is likely to result in many additional satisfying experiences as well. In general, because objects, events, states of affairs, and so on, normally have far more ramifications for us than do our mere immediate experiences of them, it makes sense to focus both our everyday preference talk and our everyday personal well-being talk upon the objects, events, and states of affairs themselves, and not upon the immediate experiences of them (that is, not on utility).

In sum: the analysis of utility being set out here is 'subjectivist' because, according to this analysis, those things to which this concept refers are certain experiences, and experiences are subjective. But the analysis of personal well-being (welfare, interests), on the other hand, is 'objectivist' because those things to which these concepts refer are, for the most part, acts, events, objects, and states of affairs, which are objective. According to this analysis, utility is a *technical* concept in terms of which our *ordinary* concept of personal well-being can, I am claiming, be better understood. So the test for whether this analysis of utility—in other words, the compromise model—is adequate is whether it does in fact enable us to understand better our ordinary concept of personal well-being. Whether the compromise model passes this test—as compared with the traditional experience model and the preference model—we shall examine next.

1.8 THE MODELS COMPARED

Now that we have the three models before us, the next step is to evaluate them. Let us begin by comparing the compromise model with its close relative, the traditional experience model that equates utility with happiness or pleasure.

First, the compromise model does not, as does the traditional model, unduly restrict what kinds of experiences count as utility. According to the compromise model, any experiences count no matter what kind, provided only that they are the objects of fully informed preferences for experiences. Thus the compromise model is less susceptible than the traditional model to the charge that it restricts what constitutes personal utility too narrowly and, being less susceptible to this charge, it thereby goes a step further in support of the liberal's ideal of toleration. Secondly, the compromise model, by relating personal utility to personal well-being simply as results are related to causes, thereby easily explains why Mary was made worse off by the stock-market crash on Black Monday even before it had affected her experiences in any way. Aside from these two differences between them, the merits and demerits of these two versions of the experience model seem to be much the same. But these two differences are significant and show the compromise model to be the superior version.

The more difficult comparison, the one that will occupy us now for the remainder of this section, is between the compromise and preference models. We have seen that both of these models are perfectly capable of explaining why the crash on Black Monday made Mary worse off before it had affected her experiences in any way. Thus the Black Monday example, as set out so far, provides no basis for choosing between them. But let us develop this example a little further. Say that, unfortunately, Mary is killed on her safari by a tiger, exactly one week after the crash, without ever having learnt about having lost her fortune. We then, typically, would no longer think of the crash as having made her worse off during this final week of her life. This is evidenced by the fact that we would then find it perfectly natural to say something such as, 'Although she died a terrible death, we can at least be thankful that she never knew she had lost her fortune.' In short: if she never became despondent, never missed out on any happiness, and so on, as a result of the crash, if, in other words, none of the expected effects upon her experiences ever occurred, or ever will occur, then, typically, we would change our minds about the crash having made her worse off. Assuming I am right about our being likely to change our minds, we now have a basis for choosing between the two models in question. Why, typically, we would change our minds cannot be explained in terms of the preference model. According to this model, Mary was made worse off, once and for all, the moment her preference for maintaining her fortune was frustrated; that she died without ever having learnt of the crash would be altogether irrelevant to the question of whether it really did cause her to be worse off during her last week of life. But why we would change our minds can be explained easily in terms of the compromise model. According to this model, prior to Mary's death, the crash very likely made her worse off. Indeed, the probability of its having made her worse off was, prior to her death, so great that we would not, at the time, even have bothered with the qualifier 'very likely'. We would have said instead simply that the crash made her worse off, its being understood that we meant worse off relative to the information available at the time (or, in other words, worse off in the sense that her 'expected' utility was lower). After Mary's death, however, it would be apparent that the crash never did, and never will, have any effect upon any of her experiences. Thus, according to the compromise model, the crash actually did not end up making her any worse off after all, which, I suggest, is exactly what, typically, a person would conclude. (Of course a die-hard exponent of the preference model might well resist this conclusion but, in doing so, could not claim the support of ordinary linguistic usage.)

In general, the preference model seems to allow too much to count as personal well-being. It required us to say Mary had been made worse off by the crash, even after we knew that she had really not been made worse off. Moreover, by allowing too much to count as personal well-being, this model precludes us from making certain important distinctions between personal well-being and what is not personal well-being. One such distinction precluded by this model is the distinction between personal well-being and self-sacrifice. Say, for example, a person prefers above all else to set himself on fire as a protest against nuclear weapons, thinking that, although it will regrettably terminate his life, such a dramatic protest might be of immense benefit to humanity. Obviously, burning himself to death for the benefit of humanity would be an act of extreme self-sacrifice. But because it would satisfy his strongest preference, the preference model commits us to saying that burning himself to death is actually conducive to his well-being. (On this, see also Sumner 1981: sect. 21; Schwartz 1982).

Although, by burning himself to death, this person (we are assuming) would indeed be satisfying his strongest preferences *in general*, he surely would not be satisfying his strongest preferences *for personal experiences*. In other words, from the domain that includes all this person's preferences, burning himself to death would, by hypothesis, satisfy stronger preferences than not doing so. But, from the more limited domain that includes only those of his fully informed preferences pertaining to his own personal experiences, I assume the opposite would be true; not burning himself to death would satisfy stronger preferences than doing so. And, according to the compromise model, it is only the satisfaction of preferences from this limited domain—that is, only the satisfaction of fully informed preferences for personal experiences—that relates to personal well-being. So, with the compromise model, a satisfactory distinction between personal well-being and self-sacrifice can easily be made. Personal well-being has to do only with what results in the satisfaction of preferences within this limited domain, whereas self-sacrifice has to do with what results in the satisfaction of preferences for the well-being of others that, as in the present case, conflict with and override the subject's strongest preference within this limited domain. Thus the compromise model allows us to conclude that, in burning himself to death for the sake of humanity, this person was indeed sacrificing himself rather than enhancing his own well-being, which is exactly what any adequate analysis must allow us to conclude. (For another distinction that the compromise, but not preference, model can handle— the distinction between preferences that have nothing to do with personal well-being and ones that do—see Haslett 1990.)

It is the experience requirement that allows the compromise model to distinguish between personal well-being and self-sacrifice more success-fully than does the preference model. But an equally important feature of the compromise model is the requirement that only preferences for experiences that are fully informed count. Strong versions of the preference model—ones that do not, as does the compromise model, incorporate a 'fully informed' requirement—fail to capture still another important distinction that the compromise model succeeds in capturing: that between the prudent and the foolish. According to strong versions of the preference model, anything that satisfies preferences must automatic-ally be deemed prudent—that is, in our best interests—no matter how foolish it really is. Thus, for example, these versions of the preference model commit us to saying such outrageous things as that the person who prefers, above all else, to begin using the drug crack is, in doing so, therefore acting in his best interests. Were all the effects on his experiences of his regularly using crack vividly apparent to him, not just the relatively short-lived highs, but the more drawn-out lows and, perhaps most important of all, the absence of those deep and meaningful experiences of life that serious drug addiction typically precludes—in other words, were his preferences for experiences fully informed—then surely he would prefer instead not to begin using crack. And since, with the compromise model, only those preferences for experiences that are fully informed count, surely it is therefore not to begin using crack that is, according to this model, in his best interests—which is, once again, exactly the conclusion that any adequate analysis must allow us to reach.

Weak versions of the preference model, ones that do incorporate a fully informed requirement of some sort, may be able to account for the distinction between the prudent and the foolish nearly as well as does the compromise model. But not even weak versions of the preference model can avoid still another problem, one raised by Richard B. Brandt (1982). Given that people's preferences change as they go through life, at what time in their lives do people have those preferences that count—that is, count for purposes of deciding which alternative maximizes their utility and thus is in their best interests? Are those preferences that count the ones people have in their childhood, or in their twenties, or in old age, or at exactly that time at which a decision as to which alternative is in their best interests is to be made, or what? It seems that no matter what time in their lives the advocate of the preference model picks as the relevant time, this model will often yield counter-intuitive results. Consider, for example, the time that most advocates of this model take to be the relevant one: the time at which the personal-welfare decision in question is to be made. Say Mary comes to you rather depressed, although only temporarily, from having broken up with John, and asks to borrow your gun so that she can shoot herself, as she much prefers death over anything else. And, significantly, this could well be her strongest preference even if her preferences were in some sense fully informed, for her depression could well, at that particular time, be exerting a stronger influence upon her preferences than even full information. But since, by hypothesis, her depression is only temporary, we may assume that it is not really in her best interests to loan her the gun. Yet if you must take her strongest preference *at that time* to be definitive, then you must conclude that loaning her the gun nevertheless is in her best interests. And you must, according to the preference model, conclude this even though you may realize that, once her depression lifts, and thus no longer has any influence upon her preferences, then her preferences will change.

The compromise model, on the other hand, avoids the change-in-preferences-over-time problem. This is because, with this model, any such changes in preferences are impossible. In short, fully informed preferences for particular *experiences* are necessarily timeless. This follows simply from what it means, according to the compromise model, for a preference for one experience over another to be fully informed. It means the preference has been influenced by perfect knowledge of what it would be like actually to have the experiences in question—that is, the preference meets the perfect-knowledge requirement. It also means the preference has not been influenced by anything else—that is, it meets the exclusory requirement as well. The exclusory requirement is the key. By its ruling out all influences on a preference other than what it would be like actually to have the experiences in question, it thereby rules out anything, such as temporary depression, that conceivably could cause the preference to change over time.

We must be very careful here not to confuse fully informed preferences for particular experiences (i.e. experience 'tokens') with closely associated preferences for experience types, preferences that clearly do change over time. Take, for example, Mr Smith, who, last 1 August, when the temperature reached 101 degrees, preferred to go swimming in his backyard pool rather than play tennis, and last 1 October, when the temperature was only 60 degrees, preferred playing tennis instead. It is accurate to say that, on 1 August, he preferred the experience of going swimming to that of playing tennis, yet on 1 October, he preferred the experience of playing tennis to that of swimming. Is it accurate also to say that his preference as between the experiences of swimming and that of playing tennis had changed from 1 August to 1 October? Yes and No. It depends upon whether the experiences in question are the 'generic' experiences of swimming and playing tennis, or the 'particular' experiences he would have had on the dates in question. Let me explain. Certainly with respect to the generic experiences in question here—that is, the experience types—his preference on 1 August obviously was different from his preference on 1 October—on 1 August he preferred the experience of swimming, whereas on 1 October he preferred the experience of playing tennis. But it is only preferences as between particular, 'dated' experiences (i.e. experience tokens) that are relevant for determining utility, and his preferences as between the *particular* experiences in question here (i.e. the ones he would have had on each of these dates) would not have changed. On 1 August, his experience of swimming would have been that of cool, refreshing relief from the heat while, in this heat, his experience of playing tennis would have been that of an unpleasant, exhausting ordeal. As between *these* two experiences, his fully informed preference on 1 August, 1 October, or any other day is, we may assume, for the cool, refreshing experience of swimming. So, on 1 October, even though Mr Smith's preference as between the experience types in question had most certainly changed, his (fully informed) preference as between the particular experiences of swimming and tennis that he would have had on 1 August had not changed. In general, although preferences between event types, and even between experience types, often change over time, fully informed preferences between particular experiences cannot change; the exclusory requirement rules out any influence upon them that possibly could cause a change. And if fully informed preferences between particular experiences cannot change then, obviously, problems having to do with preferences changing over time cannot arise with the compromise model.

This same move—that is, excluding any influence upon our preferences other than perfect knowledge—is not available to the advocate of the preference model. This is because only particular experiences have, *inherent entirely within themselves*, a basis for preferences among them (see Sect. 1.7.3 above). So only if the subjects of our preferences are particular experiences can our preferences, in principle, be based upon nothing other than perfect knowledge of these subjects themselves. But only preferences that are based upon nothing other than perfect knowledge of their subjects can be timeless. Therefore, only if the subjects of our preferences are particular experiences can our preferences be timeless. Thus, to avoid the change-in-preference-over-time problems, the advocate of the preference model would have to adopt the experience requirement, which, of course, would be to give up the ship altogether.

Finally, let us see how the compromise model compares with the preference model in avoiding that most infamous of all problems for models of utility: the problem of how valid, interpersonal comparisons of utility are to be made. With the preference model, the problem of interpersonal comparisons is serious indeed. Say, for example, that Jones prefers steak to hamburger, Smith prefers tennis to swimming, and (for some reason) we can satisfy only one of these two preferences. How exactly could we find out which would result in the most preference satisfaction? Notice that, with the preference model, it is not just that we do not have enough information to make valid interpersonal comparisons such as these, we do not even have a clear enough idea as to *how* any such comparisons would be made. That is to say, we cannot describe, or imagine to ourselves, exactly how any such comparisons would be made, even in principle. Accordingly, it is not even clear exactly what we *mean* by validly comparing different people's preference satisfaction. And if we do not even know exactly what we mean by this, then not only can we not make these comparisons with the precision that social scientists desire, but it would seem we cannot even make intelligent estimates. For, to make intelligent estimates of how one thing compares with another, surely the comparison in question must at least be meaningful.

The compromise model does not avoid the problem of interpersonal comparisons altogether. But this problem is not as serious for the compromise model as for the preference model. With the compromise model, but (as we just saw) not with the preference model, it is at least clear *in principle* what it means for one person's preference satisfaction to be greater than another's; thus with the compromise model, but not with the preference model, how one person's preference satisfaction compares with another's can at least be intelligently estimated. This is so for one reason. With the compromise model, but not with the preference model, the only kinds of preference satisfactions that count are those in the form of particular experiences. Only by having the experiences themselves can preferences for experiences be satisfied. Thus, with the compromise model, a comparison of preference satisfactions comes down to a comparison of experiences. And we already know how to make sense of comparing the experience of one person with that of another. As pointed out many years ago by John Stuart Mill (1861), we simply have qualitatively identical experiences ourselves and see which we prefer, or, if we cannot very well duplicate the experiences ourselves, we ask the person who has.

Of course no one can ever in fact duplicate, in every detail, and thus compare, precisely, the particular experiences that other people have had. But this is altogether beside the point. The point is simply that we can at least describe, or imagine to ourselves, exactly how such a comparison would, in principle, be made; thus the comparison is a meaningful one. If the comparison is a meaningful one, then we can at least go about the business of intelligently estimating it. I do not, however, know how we could compare, similarly, the preference satisfactions of different people for preferences the satisfactions of which did not all take the form of experiences. Without a common underlying element, as would be the experiences, we would then be reduced to comparing, as it were, apples with oranges; that is, comparing alternatives that are incommensurable (cf. Broome 1978: 229–32).

1.9 THE STANDARD OBJECTION

Before concluding, I shall address what has become, in one form or another, the standard objection against any experience model, which would include the one defended here. This objection is popular primarily among philosophers with intuitionist inclinations. An especially well-known variation of this type of objection is presented by Robert Nozick (1974: 42–5; see also Griffin 1986: 9–10) and goes something like this. Imagine that whatever experiences that we equate with utility were to be caused artificially. In particular, imagine that they were to be caused by an 'experience machine' attached to our brain, and that (unknown to us) our brain was floating in a tank with the electrodes from this extraordinary machine sticking into it. This little exercise in science fiction is supposed to convince us that whether something, P, is in our best interests could not be merely a function of the personal experiences P is likely to cause and prevent. There must be more to well-being than this, we are told, because if all our experiences were being artificially caused by this brain machine, then we would 'clearly' be worse off than if the very same experiences were being caused instead in the conventional way. In other words, even if how our experiences were being caused would make absolutely no difference in any of our experiences throughout all eternity, how they were being caused would nevertheless be, we are told, of great significance for our personal well-being. Exactly why, we are not told. We are told only to consult our 'intuitions', and we will see that it is so. Another favourite example of those who concoct fantastic fictions to feed our intuitions is that of being made worse off, supposedly, by having a spouse who, unknown to us, is no more than a cleverly constructed automaton, yet who, by hypothesis, will never affect any of our experiences any differently than they would have been affected had our spouse been a real person. Still another favourite example is that of our being made worse off, supposedly, by being greatly deceived in some way (such as by the unfaithfulness of our spouse) without ever learning of the deception, even though, by hypothesis once again, none of this has any effect upon any experiences we will ever have.

In reply, an advocate of an experience model need only point out one thing: the intuitions to which fantastic fictions such as these give rise are hardly reliable. Notice that the fantastic element in the last example is not that of an enormous deception being perpetrated upon us that goes undetected; what is fantastic is the hypothesis that this deception will make no difference at all in any of our experiences. In reality, an enormous deception perpetrated upon us would be bound to affect, in numerous subtle ways, our interaction with whomever was deceiving us, and therefore would probably make a major difference in our experiences. In fact, this hypothesis of no effect whatever is the most fantastic element of the floating brain and the automaton examples as well; in reality, being a floating brain, or having a spouse that was a cleverly constructed automaton, would no doubt make a difference in our experiences that was both extensive and undesirable. In reality, these things would no doubt make us very much worse off indeed. We may know perfectly well that we are to hypothesize that, contrary to reality, these things would not make a difference. Yet our emotional reactions to these things will naturally tend to reflect reality rather than this extraordinary hypothesis. And, naturally, our *intuitions* about whether these things make us worse off will tend to reflect our emotional reactions. That is why, with these kinds of cases, 'intuitions' tend not to be reliable guides. Instead of 'intuitions', what we need is a reason. We need a reason why something that could not possibly make any difference in any of our experiences would nevertheless make us worse off. Since a reason is what those who trade in these fantastic fictions never give us, I find them totally unconvincing. When confronted with these fictions, let us not allow emotions, fuelled by imagination, to get the better of intellect.

Yet let me make one thing clear. Say we had a choice between either becoming a brain floating in a tank with marvellous experiences or continuing our conventional existence but with far less marvellous experiences. About this choice, I think it correct to say the following two things. First, most of us, including myself, would choose to continue our conventional existence. Secondly, for most of us, this choice makes perfectly good sense. This, however, is what I want to make clear: the fact that it makes sense to choose our conventional existence in no way undermines the credibility of the compromise model. It makes sense to choose our conventional existence for one very good reason: our own well-being is not the only thing we value. For one thing, we value the well-being of our children and other loved ones—that is, their well-being itself—not merely our experiences of their well-being, and, as mere brains floating in a tank, we could hardly contribute much to their well-being. Of course we may value our experiences of their well-being also, which simply means that we may value their well-being for both its own sake and for our sakes. Yet, as between their well-being and our own, we may well value their well-being most (i.e. we may well prefer it). Contrary to what advocates of the preference model appear to think, however, this fact has absolutely nothing to do with whether an experience model of personal well-being is correct. In defending a version of an experience model, I need not, do not, and will not argue that it is somehow irrational, unjustified, or in some other way misguided for one to value the well-being of one's loved ones even more than one's own well-being. The contrary doctrine—that whatever one values (or prefers) most is necessarily equivalent to what is, or what one thinks is, in one's own best interests—is no more than a version of psychological egoism, a doctrine that has been discredited long ago. That we value other things, some of them even more than we value our own well-being, thus explains why, for most of us, it makes perfectly good sense to choose to continue our conventional existence, even though it results in the less preferable experiences. And that the compromise model is, after all, relevant only to one of the things we value—our own well-being—explains why the reasonableness of choosing our conventional existence in no way undermines the credibility of this model.

1.10 SOCIAL UTILITY

So far we have looked only at what, according to the compromise model, constitutes *personal* utility and *personal* well-being. Let us now complete the picture by looking briefly at what constitutes the alternative that maximizes *social* utility, and thus '*overall*' well-being. According to the compromise model, the alternative that maximizes social utility, and thus overall well-being, is the alternative that results in people having experiences that, taken together as one set, a person, whose preference was fully informed, would prefer to the set resulting from any other alternative.

For a person's preference to be 'fully informed', not only must it be influenced by perfect knowledge of all the experiences in question, but, as we have seen, it must also not be influenced by anything else (the exclusory requirement). Not being influenced by anything else assures that absence of any influences that are irrelevant. Earlier we saw that this absence of any irrelevant influences means that people's fully informed preferences at any one time in their lives must necessarily be the same as they are at any other time. But notice now that it also means that the fully informed preferences of any one person must necessarily be the same as those of any other person. Thus, for determining which alternative maximizes social utility, it does not matter whose preference is taken as definitive; all that matters, in principle, is that the preference be 'fully informed'. To emphasize this fact, let us refer to the person whose preference delineates which alternative maximizes social utility simply as 'Anyperson'. Therefore, the alternative that maximizes social utility is the one that results in people having experiences that, taken together as one set, Anyperson would prefer to the set resulting from any other alternative.

Consider a very simple case for purposes of illustration. Say that the only alternatives in question are (a) Hazel's breaking Gertrude's antique vase over Bertram's head, or (b) Hazel's continuing to endure Bertram's insults. The experiences that would result from alternative (a) are Hazel's satisfaction in silencing Bertram, Gertrude's distress over her broken antique vase, and Bertram's headache. The experiences that would result from alternative (b) are Hazel's anger from continuing to endure Bertram's insults and Bertram's enjoyment of Hazel's anger. Of the two, the alternative that maximized social utility would then be a matter of which of the following two sets, or 'bundles' Anyperson would prefer: a set consisting of Hazel's satisfaction, Gertrude's distress, and Bertram's headache; or a set consisting of Hazel's anger and Bertram's enjoyment. This preference is to be the one that Anyperson would have if she were to think of herself as having, personally, to experience exact copies of all the particular experiences resulting from whichever of the alternatives occurs. And, perhaps, we could envisage Anyperson as suffering complete amnesia immediately prior to experiencing these copies of other people's experiences. This would serve to avoid the problem of how she could experience copies that were exactly the same in every detail as the originals, while, at the same time, retaining her own personal memories (memories that otherwise would, arguably, infect these experiences so that they could not be exact copies). In any case, the 'bundle' that Anyperson would prefer delineates the alternative that would maximize social utility, and thus overall well-being.

The time has come, finally, to complete the argument against the sceptic. As between any given alternative sets of experiences, the fully informed preference of any one person must, because of the exclusory requirement, necessarily be the same as the fully informed preference of any other person. Therefore, with the compromise model, the question of which alternative maximizes social utility, and thus overall well-being, is always a question with one and only one correct answer. We may not, of course, actually know what the answer is, but there will always be an answer, and it will always be clear how—at least in principle—we would go about trying to find this answer. Thus with the compromise model, but not, I think, with any other model of utility, moral scepticism can be laid to rest.

1.11 COMPARED WITH INTUITIONIST METHODOLOGY

Throughout this chapter, I have been defending what amounts to an indirect utilitarian methodology for justifying moral norms and values. For justifying moral norms and values, this methodology appeals, ultimately, to the utilitarian standard, which ranks alternatives in terms of their conduciveness to overall well-being or utility. According to the particular version of indirect utilitarian methodology defended here, those moral norms and values that are most justified are, roughly speaking, those that would maximize overall well-being, or utility, if backed by social pressure (Sect. 1.4). The moral norms and values so justified we may refer to as those of the 'indirect utilitarian system of morality', or, simply, those of 'indirect utilitarian morality'. Since—surely—our current morality is not altogether justified in every respect, indirect utilitarian morality and our current morality will thus not be identical. But indirect utilitarian morality, no doubt, will include many moral norms and values of a sort very similar to those of our current morality—moral norms such as 'Do not steal', 'Do not lie', and 'Keep your promises', each with a few, relatively simply, easy to comply with exceptions built into them (Sect. 1.6). Let me conclude now by explaining briefly what I take to be the appropriate, indirect utilitarian methodology not for *justifying* moral norms and values such as these, but for actually *putting them to use*.

Moral norms and values are used, of course, for determining the moral status of particular acts (e.g. breaking one's promise about helping Jones study), of particular policies (e.g. never giving to charity), of particular social institutions (e.g. capitalism), and so on. Indirect utilitarian methodology for putting moral norms and values to use is usually quite straightforward. As always, the first step is determining what system of morality constitutes the appropriate frame of reference—one's 'frame of reference' being, simply, the system of morality in terms of which one's moral thinking is to proceed (Sect. 1.4). Is the appropriate frame of reference, it must be asked, current morality, indirect utilitarian morality, or what? The next step is determining, through straightforward deductive reasoning, whether or not the act, policy, or whatever, that is being examined violates one or more of the norms, or exemplifies one or more of the values, of this system. Say, for example, that the appropriate frame of reference is current morality, and that the act being examined is that of killing one's uncle so as to inherit his fortune. Current morality includes a norm that makes it very wrong, morally, to murder. Since killing one's uncle to inherit his fortune is murder, straightforward deductive reasoning reveals that the normative status of this act is, therefore, that of being, morally, very wrong.

Unfortunately, determining moral status is not always so straightforward. Say, for example, that the act in question is instead that of killing one's uncle painlessly—as he himself has requested—so as to put him out of his misery a few weeks before he dies from the extremely painful, terminal illness from which he is suffering. This is the sort of case that (I am assuming) falls within the 'borderline' area of the moral norm prohibiting murder; therefore no answer can be deduced merely from the norm itself. Norms and values in difficult cases such as this require interpretation.

According to indirect utilitarian methodology, the first step in determining which interpretation of a norm or value is most justified is to determine which interpretation leaves the norm or value most coherent with the other norms and values of whatever system of morality is being used as a frame of reference (and to determine, of course, which interpretation is most consistent with known facts). If coherence (and consistency with known facts) are not, by themselves, sufficient for settling the issue then, as a last resort, we may have to appeal to the utilitarian standard itself for 'tinkering' with, or filling up certain gaps in, the system of morality. But remember that the utilitarian standard to which we would be appealing is not a moral norm or value; it is instead a standard for justifying moral norms and values and, indeed, entire systems of morality. So in difficult, borderline cases—or, incidentally, in cases in which one questions the norms and values within one's frame of reference—morally evaluating particular acts, policies, and social institutions may require more than straightforward, deductive reasoning; more, even, then determining coherence with other norms and values and consistency with known facts. It may, as a last resort, require that we turn to the methodology for justifying moral norms and values in the first place. But only as a last resort.

This book evaluates economic systems, which are, of course, social institutions. As we have just seen, one thing common to indirect utilitarian methodology for putting moral norms and values to use for evaluating social institutions and this methodology for evaluating moral norms and values themselves is that, in either case, we are to begin by determining the appropriate frame of reference. What then shall be the frame of reference for the evaluation here of economic systems? Given my defence of indirect utilitarian methodology, the most appropriate frame of reference for me to use is that system of morality which, in terms of this methodology, is most justified—namely, indirect utilitarian morality.

The problem with this is that the specific content of a writer's ethical frame of reference should be familiar to his or her readers. But what exactly is the specific content of indirect utilitarian morality? What norms and values, exactly, are inherent in it? This is no easy question, and I do not myself claim to know the complete answer. So how then can I justify using this ideal system, rather than current morality, as my frame of reference?

I can justify it for two reasons. First, throughout this book and, in particular, in Sections 2.2 and 3.2, I shall be setting out, in general outline, what I take the content of indirect utilitarian morality to be, or at least that part most relevant for my purposes here. Secondly, except in so far as I indicate otherwise in this general outline, I shall simply assume (tentatively) that indirect utilitarian morality is identical to current morality. Since, in this general outline, I do not try to set out indirect utilitarian morality in any great detail, and since what I do set out bears a striking resemblance to current morality anyway, what this assumption of identity means is that, for all practical purposes, the frame of reference for my evaluation of economic systems shall mainly be current morality—that is, the morality found in present-day, large industrialized societies.

Do not misunderstand me: by tentatively assuming an identity between indirect utilitarian and current morality (except in so far as I indicate otherwise), I am certainly not claiming that, to this extent, the two moralities are in fact identical. They are not. But the difficult journey from current morality to ideal morality (which is what I take indirect utilitarian morality to be) never ends and, all along the way, moral decisions continue to have to be made. A good analogy for this journey is that of the ship which must be rebuilt while remaining at sea. If the ship is slowly rebuilt part by part, then, over time, a whole new ship emerges, even though it has remained at sea all the while. Likewise, I suggest, over time a new, more ideal morality will slowly emerge; yet all the while we must make moral decisions by continuing to use, for the most part, whatever moral norms and values society currently has (that is, all the while we must 'remain at sea'). Tentatively assuming an identity between current and ideal morality merely reflects this need for continuing to use current norms and values while, over time, ideal morality slowly emerges.

Finally, it is worth noting how the indirect utilitarian methodology I shall be using here for evaluating economic systems compares with the moral methodology that, I think, is used most often these days: intuitionist, or reflective equilibrium methodology (for this methodology, see Rawls 1971; see also Haslett 1987*b*). Just as do practitioners of intuitionist methodology, I shall, for the most part, be taking current morality (as evidenced by our moral 'intuitions') as my starting-point for evaluating alternatives. And, just as do practitioners of intuitionist methodology, I shall try to determine which alternative is most justified in terms of this starting-point by appealing largely to coherence and consistency with known facts. Thus, as we see, intuitionist methodology and this indirect utilitarian methodology are very similar. This similarity is significant. It is usually assumed that intuitionist and utilitarian methodologies are altogether incompatible, but there is no serious incompatibility between intuitionist methodology and the indirect utilitarian methodology I shall be using here. Moral intuitionists, normally adamant opponents of utilitarianism, should thus be reasonably comfortable with this methodology. And they may, therefore, find themselves in agreement with the conclusions reached.

Where indirect utilitarian and intuitionist methodologies do, however, differ is that, with intuitionist methodology, there is no final criterion for justifying moral norms and values that is equivalent to the utilitarian standard. Though it will usually make little difference in practice, this difference is absolutely crucial in theory. Without something equivalent to the utilitarian standard to serve as a criterion of ultimate appeal, intuitionist methodology is left without any final means of adjudicating between people with radically differing moral 'intuitions'. Thus intuitionist methodology is forever at the mercy of moral sceptics, those who claim there is no correct answer, and who thereby undermine all answers, no matter how well supported they may appear to be. With indirect utilitarian methodology, on the other hand, if all else fails there is always the utilitarian standard; so there will always be a correct answer in principle. Thus, with this methodology, moral scepticism can indeed be laid to rest, and the search for answers can proceed with confidence.

2
LIBERTARIANISM

2.1 INTRODUCTION

Indirect utilitarianism is the view that the appropriate standard for evaluating, and thus justifying, normative criteria—norms, values, and standards—is the utilitarian standard. But, according to indirect utilitarianism, the utilitarian standard is not the appropriate standard for evaluating the various things (aside from other normative criteria) that are typically evaluated in terms of normative criteria—things such as acts, policies, and social institutions. A lawyer whose practice consists of giving advice not to laymen, as most lawyers do, but to other lawyers who are working on unusually difficult cases is known as a 'lawyer's lawyer'. Similarly, the utilitarian standard is, according to indirect utilitarianism, a 'normative criterion's criterion'.

What it means for normative criteria to be justified in terms of the utilitarian standard is that they are to be justified in terms of how conducive they are to overall well-being, or utility, with everyone's well-being, or utility, being given equal consideration. And, according to the particular indirect utilitarian methodology that I advocate, the focus is to be upon how conducive to utility the alternative criteria being evaluated are *if backed by social pressure*. Focusing upon their conduciveness to utility if backed by social pressure is, as explained in Chapter 1, crucial; in determining conduciveness to utility, this focus allows human fallibility to be taken fully into account.

This methodology is designed to be used for justifying the norms and values of both personal and political morality. Personal morality consists of those norms, values, and standards designed for, or applicable to, the decisions of individuals, while political morality consists of those norms, values, and standards applicable to the decisions of political units—nations, states, cities, and so on—or their governments. Since the choice of an economic system is a political decision, it is in terms of what I am calling 'political' morality that alternative economic systems are to be evaluated.

In this chapter, I shall do three things. First, I shall sketch indirect utilitarian political morality. Since, as explained earlier (Sect. 1.11), I am taking indirect utilitarian political morality as my ethical frame of reference for evaluating alternative economic systems here, it is essential that I set out, at least in outline, what I take this morality to be. As also explained earlier, I am (merely for purposes of argument) assuming that, in so far as this outline does not indicate otherwise, the particular norms, values, and standards of indirect utilitarian morality are identical with those of our current political morality.

Next, I shall in this chapter sketch a leading competitor of indirect utilitarian morality—namely, libertarian morality. A consideration of libertarian morality here is important. First, versions of libertarian morality are widely, and adamantly, defended these days. Secondly, this morality provides the main philosophic rationale for political conservatism in general, and for a conservative economic system in particular, that of *laissez-faire* capitalism. By 'capitalism', I mean, roughly, an economic system characterized largely by private ownership of the means of production—that is, capital—and by an absence of central economic planning. And '*laissez-faire*' capitalism is capitalism without any governmental interference in economic transactions at all, beyond that of merely upholding people's property rights and enforcing contracts. Any serious evaluation of *laissez-faire* capitalism must take the form of an evaluation of libertarian morality, which constitutes its philosophic basis.

Libertarian morality is premissed upon a rejection of utilitarian morality of any kind. The fundamental objection that libertarians have against utilitarian morality—an objection they are not alone in making—is that utilitarian morality does not take the difference between persons seriously; that it calls only for a maximization of utility, as if everyone's consciousness could be reduced to just a single consciousness. In reality, the libertarian objects, we are many, irreducibly separate persons, whose rights and interests must never be sacrificed in order to maximize utility. This then brings me to the third thing I shall do in this chapter: try to answer this fundamental objection against utilitarianism.

Overall I shall argue that, as compared with indirect utilitarian morality, libertarian morality is hopelessly deficient and, consequently, so is the extreme version of capitalism that follows from it.

2.2 INDIRECT UTILITARIAN POLITICAL MORALITY

What I mean by 'indirect utilitarian' morality is that morality justifiable in terms of the indirect utilitarian methodology set out here (Sect. 2.1 and esp. Ch. 1). Earlier I outlined, in very general terms, what I think indirect utilitarian personal morality is like (Sect. 1.6). Let me now outline what I think indirect utilitarian political morality is like, political morality being, once again, that morality in terms of which governmental acts and policies, laws and regulations, and social institutions are to be evaluated.

The most general standard of this political morality—the general welfare—is a variation of the utilitarian standard. According to this general standard of political morality, governmental decisions are to be evaluated in terms of how conducive they are to the well-being (or 'utility') of those currently within the government's jurisdiction, with everyone's well-being, of course, to be given equal consideration (for the definitions of 'well-being' and 'utility', see Sects. 1.7 and 1.10). This focus upon those currently within its jurisdiction certainly does not mean that a government should not, in cases of famine or other special need, help those outside its jurisdiction. Such aid is commendable. But the primary concern of a government, I am claiming, should be those over whom it actually governs. And, from among those over whom it governs, the government should focus not upon the welfare of any special interest group, but instead upon the *general* welfare. All of this, I think, corresponds well with most people's moral 'intuitions' about a government's appropriate domain of concern and can be justified in terms of the utilitarian standard. This justification would, in part, appeal to the efficiency of a division of governmental labour in which each governmental unit is to be concerned primarily with that which it 'knows' best; namely, its own constituency.

The general welfare, however, represents a very abstract ideal indeed. The very fact that it is so abstract often makes it extremely difficult for public officials, fallible as they (like any human beings) are, to know exactly what this ideal calls for. But, as already pointed out, indirect utilitarian methodology is specifically designed so that human fallibility is to be taken fully into account in justifying a morality. Any morality so justified will, therefore, be one that compensates, somehow, for this fallibility. Indirect utilitarian political morality, I claim, compensates in two ways. First, this morality features a number of values, or ideals, that are more specific than the general welfare, and that therefore serve to *guide* public officials (or anyone else) in their pursuit of this abstract ideal. Secondly, this morality features a number of rights (against the government) and other moral norms that are always to be given priority over the general welfare, and that therefore serve to *constrain* public officials (or anyone else) in their pursuit of it. These constraints help assure that public officials do not, through well-meant but misguided interpretations of the general welfare, end up doing more harm than good.

Consider first the values, or ideals. As I see it, among the most significant are freedom, productivity, and two ideals of justice: equal opportunity and equal access to the basic necessities of life. Other, even more specific, ideals of indirect utilitarian morality include health, safety, recreation, education, and a clean environment. Ideas such as these serve as guides to the general welfare in the following sense: we are to assume that, other things being equal, the more any one of them is realized, the greater the general welfare.

Of course other things rarely are equal; increases in one thing of value often entail decreases in other things of value. These increases and decreases must then be weighed against one another and an attempt made to find the best possible compromise or trade-off. So none of these values or ideals are to be taken as absolute; they are all to be viewed as susceptible to being overridden in conflicts with other values or ideals. They are all, in other words, only prima facie. And, according to indirect utilitarian morality, that standard to which, in the end, we are to appeal in trying to find the best possible compromise among them is, of course, that of the general welfare. There is no circularity in using these values or ideas to guide us in determining what, other things being equal, is in the general welfare, and then, as a last resort, using the general welfare to find the best possible compromise when other things are not equal, when, in other words, faced with 'conflicting' values and ideals.

Consider next the *norms* of political morality, including rights against the government. Among the norms of political morality are ones similar to those of personal morality, except that they are applicable to governments rather than persons. They are norms requiring, for example, that debts be paid, and that communication be truthful. The rights against the government include freedom of speech, freedom of religion, and freedom from any discrimination on the basis of race, sex, religion, or national origin. While the values and ideals of indirect utilitarian political morality, which include among them the general welfare, delineate goals to be realized to as great an extent as possible, these rights and other norms delineate those things a government must not do in attempting to realize these goals. They are 'side constraints' upon governmental activity. As I mentioned above, the indirect utilitarian rationale for these rights is human fallibility, the fallibility, in particular, of government officials. Because government officials, in restricting people's speech, religion, and the other things protected by these rights, are especially subject to rationalizations and miscalculations, and because it is not often that government restrictions on these things maximize utility anyway, it is 'best'—i.e. it maximizes 'expected' utility—for governments to be precluded altogether from restricting these things. (For a more complete statement of the rationale for these rights, see Haslett 1987a; see also Sumner 1987.)

Incidentally, just because these rights thus have a consequentialist justification, they are not therefore mere consequentialist 'rules of thumb'. They are not, in other words, mere heuristic devices that are to be used only in so far as people are unable (because of human fallibility) to apply some more fundamental consequentialist moral principle. The critics of indirect utilitarian morality seem forever incapable of grasping one simple fact: if indirect utilitarian morality is properly formulated, the rights and other moral norms of the code are themselves the norms of morality; no 'deeper' consequentialist moral principle is hovering in the background, waiting to override these rights whenever the veil of human fallibility lifts. This is not to say that indirect utilitarian morality rules out entirely any appeals to consequences—no matter how dire the emergency—in overriding what these rights ordinarily prescribe. But the legitimacy of any such appeals is built into these rights themselves, as potential exceptions the scopes of which are carefully limited, thereby precluding any general 'collapse' into consequentialism or direct utilitarianism. (For further elaboration, see Haslett 1987*a*: esp. 125–35.)

Two rights against the government that, I submit, are part of indirect utilitarian morality deserve special mention. The first is a right to quit one's job whenever one wants—in other words, a right not to be enslaved. I think it obvious enough that circumstances in which slavery is in the general welfare are, to say the least, rare, and that government officials cannot detect these rare circumstances accurately enough ever to justify taking a 'chance' on slavery. Thus, according to indirect utilitarian morality, the government itself may not make slaves of people, nor may it enforce private contracts of slavery.

The second indirect utilitarian right against the government that deserves special mention is a right to life, a right that prohibits the government from ever deliberately killing innocent people against their will. Like all indirect utilitarian rights, the right to life serves as an absolute side constraint upon the government's pursuit of any goals, including the general welfare. This right follows from the fact that, from an indirect utilitarian perspective, any law, decision, or policy that allows public officials deliberately to kill innocent people against their will is always a bad 'general welfare' gamble.

Take, for example, the 'survival lottery' law proposed by John Harris (1975). This law would allow the government to kill healthy people, randomly chosen by lottery, and then distribute their organs to those in need of organ transplants so that, for every one person thus killed, at least two people in need of transplants will be saved. Since this law would appear to bring about a net gain in the number of people who survive, it may not be obvious at first why even it is a bad 'general welfare' gamble. Yet it is. In the first place, people who had deliberately lived healthy lives might, in spite of their best efforts, nevertheless end up being killed for the sake of people who had deliberately lived unhealthy lives through smoking, drinking, eating too much, exercising too little, and so on. To this extent, therefore, people would lose control over their own lives. The anxiety resulting from this very visible loss of control would, by itself, probably outweigh any gains in survival there might be from such a law. Most people who buy state lottery tickets these days feel as if they may well be among the winners, even though the odds are overwhelmingly against it; likewise, most people would probably feel that, in the next survival lottery, those chosen to be killed might well include them, even though the odds were overwhelmingly against it. This seems to be how human psychology works. Moreover, and more serious yet, by the government's guaranteeing those who smoke, eat, or drink too much, exercise too little, and so on, that they will get whatever transplants they may as a result need, the government would thereby undermine people's incentives to remain healthy on their own. And the result of this might well be less survival overall even in spite of the survival lottery. There is, after all, some truth to Charles Murray's Law of Unintended Rewards, which states that 'any social transfer increases the net value of being in the condition that prompted the transfer' (1984). So, aside from the anxiety it would create, which would be bad enough, a survival lottery might even be self-defeating, that is, end up decreasing the very thing—survival—that it is designed to increase. (Harris does stipulate that an exception to those eligible for lottery benefits is to be made to those who bring their misfortunes on themselves. But if this exception were to be interpreted leniently, so that it exempted from lottery benefits only those guilty of abusing their bodies in the most blatant ways possible, then this exception would provide very little incentive for the average person to take care of himself in any way beyond that of merely refraining from the most blatant of bodily abuses. If, on the other hand, this exception were interpreted strictly, so that it exempted from lottery benefits anyone who smoked, drank, overate, underate, exercised too little, or did any of the many other things generally considered unhealthy even though not the most 'blatant' of bodily abuses, then it might well be more difficult to find eligible lottery recipients than it currently is to find organ donors, thereby nullifying the whole scheme.)

I could go on with the disadvantages of a survival lottery, going into the dangerous precedent it would represent, the line-drawing problems it would create, the justifiable norms of justice it would undermine, the potential for administrative abuse, and so on, but perhaps enough has been said. A law such as this, in spite of initial appearances to the contrary, is very unlikely indeed to be welfare maximizing. Even more counter-productive would be for, say, the FBI to take it upon themselves to roam about the country secretly snuffing out the lives of those it judged for some reason to be 'undesirable', and thus a threat to the general welfare. No individuals could possibly be trusted to make such judgements accurately enough for a practice like this to come even close to being welfare maximizing. In sum: it is all too rare for any laws, practices, or actions that violate the indirect utilitarian right to life actually to be in the general welfare—and fallible government officials are all too subject to miscalculations in trying to pick out the few exceptions—for any such laws, practices, or actions ever to be good 'general welfare' gambles. Thus, with indirect utilitarian morality, all such gambles are, by means of this right, prohibited. (For more detail, see Haslett 1987*a*: ch. 7.)

To conclude this general description of indirect utilitarian political morality, I must introduce into the discussion an important, though controversial distinction: that between 'negative' and 'positive' rights. Negative rights are rights *not to be harmed* in some way, such as, for example, by being killed or by having one's property stolen. Positive rights, on the other hand, are rights to some *benefit*, such as, for example, adequate medical care or some minimum standard of living. The clarity of this distinction has often been questioned, and with some justification. But we need not pursue this debate here; for present purposes this rough distinction between rights not to be made worse off in some way, and rights to be made better off in some way, is clear enough.

The indirect utilitarian rights we have been discussing so far—ones that preclude the government from restricting certain freedoms, or harming us in certain other ways—are all negative rights. Even the right to life just discussed is to be interpreted as a negative right—the right not to be killed. And, as we have seen, the general rationale for these negative rights is that they preclude certain government 'general welfare', or 'utility', calculations, ones that should be precluded in that they inevitably give rise to bad 'general welfare' gambles. A similar rationale may be possible for certain *positive* rights against the government as well, such as a right to a minimum standard of health care, and a right to a certain level of education. The rationale would be that (1) the likelihood that governmental provision of these benefits would be in the general welfare, and (2) the likelihood that government officials would be mistaken in trying to pick out the exceptional circumstances in which providing these benefits would not be in the general welfare, are, in combination, such as to make it 'best' that these benefits simply be made a matter of right (thereby precluding any governmental 'utility' calculations regarding whether or not to provide the benefits).

But, although I am confident about the case for negative rights and other norms of political morality, I am less confident about the case for positive rights. The main problem is that positive rights, contrary to negative rights, presuppose the solution to enormous line-drawing problems. For example, how much health care, exactly, should a right to a 'decent minimum' of health care include? Does this vary with the resources available in the country in question? How much in the way of resources would justify how much of a right? Also positive rights, but rarely negative rights, give rise to conflict-of-rights problems—ones closely related to these line-drawing problems. If, for example, the positive right to a decent minimum of health care conflicts with the positive right to some specified degree of education—since resources are insufficient for the full realization of both rights—which should be given priority?

Rather than relying upon positive rights for the justification of governmental benefits, an alternative approach—one that I suspect is more justifiable from an indirect utilitarian perspective—is to rely, for their justification, upon those values and ideals that are part of ideal utilitarian morality—values and ideals such as equal opportunity, freedom, security, health, education, and the general welfare, values and ideals the most extensive possible realization of which constitutes (according to indirect utilitarian morality) the goal of governments. Clearly government health benefits, education benefits, and other benefits often said to be a matter of positive right, can be justified as well by appeal to these values and ideals. And by thus bypassing positive rights, we therefore bypass the line-drawing problems that must be solved in formulating these rights. Of course with the 'values' approach to the justification of governmental benefits, lines must be drawn for determining how much realization of each value, *vis-à-vis* the realization of competing values, is in the general welfare. The point, however, is that, with the 'rights' approach, it would appear that, merely in order for the rights in question to be well formulated, these lines must, more or less, be drawn from the beginning. With the values approach, on the other hand, these lines need not, from the beginning, be drawn merely for purposes of formulating the values in question. These values can stand on their own simply as prima facie, without predetermined boundaries. Therefore this drawing of lines can be a constant, on-going legislative process of trial and error, and of adaptation to ever-changing circumstances. For largely these reasons, I shall not take the rights approach to the justification of any governmental benefits here, but shall take the values approach instead.

2.3 LIBERTARIAN MORALITY

Having sketched indirect utilitarian political morality, let me now, for purposes of comparison, sketch libertarian morality. The libertarian appears to be motivated by the quest of reducing morality to its bare essentials, to its most fundamental principles—principles that, for the libertarian, take the form of universal moral rights. And, with libertarian morality, we need not make a distinction between personal and political morality, for the same set of fundamental, libertarian rights is said to be applicable to all decision-making, whether personal or governmental. Libertarians disagree about what exactly these fundamental rights are, but the general idea is that these rights are to protect us from any violations against our liberty, bodies, property, and contracts. The seventeenth-century philosopher, John Locke, can be credited with being the precursor of contemporary libertarianism, and Robert Nozick is its most influential contemporary exponent. These two shall represent libertarian thought for purposes of the discussion here. (Note: I shall be focusing on Nozick's views only as set out in *Anarchy, State, and Utopia*. Although Nozick has recently expressed second thoughts about some of the views he propounds in this book (Nozick 1989: 286–7), *Anarchy, State, and Utopia* is likely nevertheless to remain the most influential libertarian treatise for many years.) The fundamental rights recognized by Locke (1690), and endorsed by Nozick (1974: 10), are six: life, liberty, health, property, the right to defend ourselves against violations of these rights, and the right to punish transgressors against these rights. And the right of property is usually interpreted as protecting us not just against theft, but against fraud and breach of contract as well. Other libertarians will, of course, defend slightly different lists of fundamental rights, but all are, I think, close enough to Locke for his list to serve as representative of this philosophy.

According to Locke, we voluntarily 'transfer' our rights to defence and punishment to the government, along with as much of our right to property as the government needs in order to carry out its legitimate functions. The remainder of our rights we retain, and no one, not even the government, may violate them. The legitimate functions of the government are delineated by the rights we have transferred to it and those we have retained. The government is thus to defend us against violations of our rights, punish wrongdoers, do whatever may be necessary for carrying out these functions, such as establish an army, a police force, jails, system of courts, and tax us enough to cover the necessary costs of these things. But the government may not do anything beyond these minimal activities, or tax us one cent more than is necessary for covering the costs of these activities; doing anything else would necessarily violate one of the rights the people have retained, these rights being ones to life, liberty, health, and property. A state in which government activity is strictly limited in this way is approvingly referred to by Nozick as a 'night watchman' state (1974: 25–7). Some libertarians go so far even as to hold that any government activity of any sort is illegitimate; that all activities, including defence and punishment, should be carried out privately through voluntary associations (see e.g. D. Friedman 1973). I shall, however, focus here on the more moderate form of libertarianism advocated by Locke and Nozick.

The key to understanding exactly how much constraint even this more moderate form of libertarianism places upon government activity is this: our fundamental rights, according to the libertarian, are all 'negative', not 'positive', ones; they are not rights to benefits, but are rights not to be harmed. Think of it this way. We are all currently at some level of well-being with respect to our liberty, health, and property—some of us at very high levels, some at very low levels. Locke's rights to liberty, health, and property, being negative, forbid anyone from doing anything to us that puts us anywhere below the levels that we are currently at—unless, that is, we have (explicitly, or implicitly) consented to it. But, by the same token, these rights do not require anyone, including the government, to do anything to put us anywhere above our current levels, no matter how low these levels and thus our need for help may be. So, for example, the right to property forbids us from taking anything from the starving pauper but does not require us to give him a single cent. Likewise, the starving pauper's right to life, being a negative right, forbids us from killing him, but does not require us to do a single thing to keep him alive. Of course the libertarian will be quick to point out that it is certainly to be hoped that people will voluntarily do what is necessary for keeping him alive. But his rights to life, liberty, property, and health do not require that anyone do so.

Consider now how much these rights, being negative, constrain government activity. Although private individuals are not required to do anything for anyone else, such as feed the starving pauper, they are certainly allowed to do so. But the libertarian government is not even allowed to do so. This is because anything the government could do would require the expenditure of some tax revenue, but the libertarian government is strictly forbidden from collecting or spending any tax revenue for purposes other than preventing and punishing violations of libertarian rights, and (since they are all negative) none of these rights would ever be violated by the mere failure to provide food for paupers or any other benefits. So any use of tax revenue for purposes such as feeding paupers would be illegitimate, since it would amount to a violation of the property rights of those who had been forced to pay the taxes used for these purposes. It would amount to no less than outright governmental theft.

This would be true even though the government had been democratically elected. Nothing short of voluntary consent from *everyone* would allow the government to use tax revenue for any purposes other than defence or punishment. And we know that unanimous consent is impossible. This does not, of course, prevent those individuals who want to help feed paupers from doing so on their own or through private agencies. But the libertarian government must play no role at all in this charitable effort, not even that of collection agency and co-ordinator for voluntary donations, for even mere collection and co-ordination would require some tax expenditure.

What economic system libertarian morality entails is thus clear: *laissez-faire* capitalism. The government's role in the economy is strictly limited to making sure that businesspeople do not renege on their contracts, defraud anyone or, in any other way, violate anyone's negative rights to life, liberty, health, or property; government may not restrict businesspeople's freedom of activity for any other purposes whatever; to do so would be a violation of their right to liberty or to property, or both. This means no regulations (beyond the bare minimum necessary for protecting us against a violation of our rights), no tariffs, no quotas, no subsidies, no minimum wages, no mandatory price ceilings or floors, no redistribution of income, and no central economic planning of any sort. So the libertarian is the arch-conservative, the most adamant of all defenders of the free market, of the 'invisible hand' of supply and demand.

There is something very attractive about libertarian morality. Part of the attractiveness is, I think, its simplicity. It reduces the vast complexities of personal and political morality to just a few easily understood rights. To be sure, sophisticated libertarians will admit a little more to the libertarian code of morality than just these rights; they will, in particular, admit virtues, such as freedom, productivity, and peace. Nevertheless, just these few, simple rights, which have priority over everything else, remain at the heart of libertarian morality.

But perhaps the most attractive feature of libertarian morality is its 'live and let live' spirit. The libertarian does not want to do anyone any harm; libertarian rights forbid it. The libertarian wants only to be left alone to live his or her own life as he or she sees fit. What, after all, could be more reasonable than this? Indeed, the libertarian will say, if only everyone were to be this reasonable, what a wonderful world it would be. No one would ever deliberately harm anyone else. War and crime would be extinct. All would be free to do as they please. Moreover, although people would not be required to help the needy, they would certainly be encouraged to do so of their own free will. And did I not myself suggest in Section 1.5 that encouraging charity by means of making it a virtue was preferable to requiring it by a norm of personal morality?

Unfortunately, in spite of the ease with which libertarian morality seems able to capture the imagination and stir the emotions, it, and the *laissez-faire* version of capitalism derivable from it, are totally and hopelessly inadequate. Let me try to explain why.

2.4 THE MORALITIES COMPARED

The first, and most important, thing to keep in mind about libertarian morality is that neither Locke nor Nozick has provided us with a serious, non-circular *argument* in justification of this morality. Jan Narveson (1988) attempts to justify libertarian morality in terms of self-interest, but this is a justification that, I argue below (Sect. 2.6), could never work. Nozick hints that an argument in justification of libertarian morality can be given by relating this morality to what is necessary for a meaningful life. Yet, as of now, the argument remains unstated.

Even in spite of this glaring lack of argumentative support for it, many people today take libertarian morality very seriously. Why? The answer can only be that they think this morality is supported by our basic moral 'intuitions'.

But, given that people's basic moral intuitions differ, mere intuitive support for a morality is weak support indeed. It leaves us with no means of adjudicating between these differences in people's basic moral intuitions, and thus leaves us at the mercy of the moral sceptic. Thus moral intuitions, even ones common to most people, can never be definitive of what morality is most justified.

Yet moral intuitions, if they are ones common to most people, can be important clues to what morality is most justified. These intuitions typically reflect society's most deep-rooted moral beliefs, ones that, having survived the test of time, may well therefore be justified after all. So if the moral intuitions common to most people do support the morality one is advocating, then this is at least enough to shift the burden of proof upon those who oppose this morality. This, I suspect, is what most advocates of libertarian morality are counting on: enough intuitive support for their morality so as to shift the burden of providing argumentative support upon their opponents. I shall try to show in what follows, however, that not only is libertarian morality seriously lacking in argumentative support as already pointed out, but, as compared with indirect utilitarian morality, it is seriously lacking in intuitive support as well, thus leaving it with no real support at all.

Let us begin by a closer look at what sorts of things libertarian morality will allow, and not allow, governments to do. The sorts of things governments in large, industrialized societies today typically do can be classified (very roughly) as follows. First, they protect us against, and (through the judiciary) provide the means for seeking relief from, 'individual' harms committed by other human beings—harms such as those of theft, assault and battery, fraud, and breach of contract. (I assume, incidentally, that any governmental punishment of those unlawfully inflicting these harms falls under the category of 'protecting' us against them.)

By 'individual harms', I do not mean harms *to* individuals, but harms *by* individuals. Individual harms are brought about by a person, or a group, acting individually. Paradigmatic examples are that of a person punching someone's nose, or stealing someone's car. But the 'individual' causing the harm can be a group too. I am using the word 'group' here in a broad, somewhat technical sense, in which it refers to any two or more people acting together, co-operatively, with some feelings of solidarity and common leadership. Group behaviour is to be distinguished from mere collective, or joint, behaviour, which is the behaviour of people, or groups, acting independently of one another, without any feelings of solidarity or common leadership. Notice that a group, in this broad, somewhat technical sense, can be anything from a relatively permanent, highly structured organization, such as a corporation, to a temporary, unstructured group banded together for a single act, such as a lynching mob (even a lynching mob will normally be characterized by at least rudimentary feelings of solidarity and common leadership). The individual harms caused by corporations might include such harms as the *Valdez* oil spill in Alaska, or the death of employees by means of the corporation's deliberately exposing them to deadly chemicals without their consent. And, of course, the individual harm caused by a lynching mob will be an unlawful hanging. The extent of the responsibility of those people within a group for the (individual) harms the group causes depends upon their position, or degree of leadership, within the group, and upon how much control over the harm they had.

Contrasted with individual harms are 'collective' harms. Collective harms differ from individual harms in that they are not caused by a single person or group, but by the behaviour of a number of people or groups acting independently of one another, without feelings of solidarity or common leadership. Paradigmatic examples of collective harms are the harms from pollution or overcrowding. Collective harms are, in other words, the cumulative result of many acts by many independent agents. Typically, no one of these acts will be 'individually' harmful, only their cumulative result will be harmful (cf. Buchanan 1985: 68). I say 'typically' because those acts contributing to a collective harm will sometimes include some acts that are individually harmful. But a collective harm is such that, even if the relatively few individually harmful acts contributing to it were eliminated, serious harm would still remain.

As we have seen, the version of libertarianism being considered here allows governmental protection against, and only against, violations of the rights recognized by libertarian morality. And this version of libertarianism recognizes rights to life, health, liberty, and property. Thus, at the very least, this version of libertarianism allows governmental protection against individual harms to our life, health, liberty, and property. It is clear, therefore, that it allows governmental activity falling within the general category of 'protection against individual harms'. So, of course, does indirect utilitarianism, although the two views differ somewhat in what individual harms they protect against.

But governments today protect against, and provide relief from, not only individual harms, but collective harms as well. Among the collective harms that governments today typically provide relief from are ones resulting from innumerable market transactions, none, or only few, of which may be harmful when considered in isolation, but the totality of which have cumulative results that are very harmful indeed. One of the cumulative results of market transactions is that some people (those whom the market leaves relatively poor) lose political power to other people (those whom the market makes relatively rich). The poor, of course, retain a vote just like the rich, but they lose, to the rich, important means of financially influencing candidates, governmental officials, and public opinion. So, although the poor retain full *de jure* political power, they lose important *de facto* political power, a loss incompatible with the ideals of democracy. This is one collective harm that governments today, which are committed to democratic ideals, attempt to remedy. These remedies include, among other things, certain restrictions upon political contributions (and, of course, upon outright political bribery), and upon lobbying public officials.

Another collective harm that is the cumulative result of innumerable market transactions is a dramatic decrease in opportunities for the children of those whom market transactions leave poor. These children, through absolutely no fault of their own, suffer from far fewer opportunities than the children of those whom these transactions make rich. Governmental remedies against this harm typically include various educational, medical, and nutritional subsidies that help provide the children of those left poor with more opportunities.

Perhaps the most obvious collective harm that governments today seek to protect us against is that from pollution of various sorts. We might also include governmental unemployment compensation under this category of protection against, and relief from, harms that are the cumulative result of innumerable market transactions—at least to the extent that the unemployment compensated for came about simply through a downturn in the economy. Another collective harm that governments typically protect us against is that resulting from unregulated commercial and industrial development. No single, new commercial or industrial project may be harmful considered individually, yet the cumulative result of numerous such projects may turn otherwise pleasant residential areas into nightmares of ugliness, noise, and congestion. The typical remedies for this type of collective harm are, of course, zoning regulations. In general, governmental protection against what economists call 'negative externalities' (Sect. 3.1) often falls under this category of governmental activity (i.e. the category of protection against collective harms).

Yet all governmental activities falling under this category are forbidden by libertarian morality or are, at best, problematic, since libertarian rights, as they are generally interpreted, offer protection only against individual harms. One does not have a right not to be harmed by, for example, lightning, since there is no identifiable agent that has control over this harm (leaving aside God, against whom we presumably have no rights). As it is with lightning (so the typical libertarian seems to believe) so it is with collective harms: no agent has control over a collective harm as a whole. To be sure, each of the agents contributing to the collective harm has control over its particular contribution to the harm (which, by itself, causes no harm). But, since a collective harm is the cumulative result of a number of contributions by a number of agents each acting independently, none of them has control over the harm as a whole. And, once again, if there is no identifiable agent with *control* over a harm, then the harm violates no libertarian rights; thus collective harms violate no libertarian rights. So if collective harms violate no libertarian rights, then, as we have seen, any governmental activities designed to protect people against these harms are beyond what, for libertarians, is the legitimate scope of government.

Furthermore, many collective harms resulting from market transactions are, so the libertarian is likely to claim, best interpreted as resulting from mere failures to act rather than from acts. The idea here is that these harms reflect the fact that there are those who have not been chosen to participate in any reasonably profitable market transactions (they have not been offered a decent job, for example), and not choosing someone for something is a failure to act, rather than an act. But to the extent that collective harms are indeed the result of failures to act, then libertarian morality clearly offers no protection against them because libertarian rights (being negative) protect only against harmful acts, not harmful failures to act.

The political morality of indirect utilitarianism, on the other hand, has no particular difficulty justifying governmental activities designed to provide relief from collective harms. According to this political morality, harms need not violate any moral rights before the government is justified in doing something about them. All that is necessary for the government to be justified in doing something about them is for doing something to be in the general welfare (and not violate any indirect utilitarian rights against the government). Since, as is perfectly obvious, governmental relief from collective harms is indeed often in the general welfare, indirect utilitarian morality has no difficulty justifying such relief. I conclude, therefore, that indirect utilitarian morality is more conducive than libertarian morality to the values and ideals that relief from collective harms helps realize, values and ideals such as those of democracy, justice, equal opportunity, and protection of the environment.

Nozick does suggest one form of relief against the collective harm of pollution that, he thinks, is compatible with libertarian morality. He says that anyone who has been harmed by pollution can institute, in court, a civil, class action against the polluters for restitution, with each polluter being liable only for that proportion of the restitution equivalent exactly to that proportion of the collective harm specifically attributable to his or her pollution (1974: 79–81). Why libertarian morality should allow even this much relief is not altogether clear. If, even though none of the individual acts that result in collective harms violate anyone's libertarian rights, these acts count as violations of libertarian rights anyway, then many of the kinds of governmental activities traditionally held to be illegitimate by libertarians may become legitimate as protection against these kinds of violations. I have in mind here governmental activities such as welfare, which provides protection against the collective harm of reduced opportunities for children of the poor. If, on the other hand, the individual acts that result in collective harms do not count as violations of libertarian rights, then why a person is entitled to restitution for collective harms cannot be explained in terms of libertarian morality, since, with libertarian morality, a person is entitled to restitution only for acts that do violate his or her rights. But, assuming that Nozick were to find a satisfactory way out of this dilemma, class actions would hardly be satisfactory remedies for the harms of pollution anyway. In the first place, how could the exact proportion of the harm specifically attributable to any one polluter ever be determined precisely enough? And with, say, lung cancer, even though we can be fairly sure that pollution will cause a certain number of these cancers overall, how could we ever be sure that, in any given case, the cause was pollution rather than something else? In the second place, the only really satisfactory remedy for pollution is somehow to control it so that people do not get the lung cancer, or whatever, to begin with; mere restitution after people have already contracted the disease is too little, too late. And how are we ever to extract adequate restitution from polluters for, say, permanently destroying the earth's ozone layer? All the money in the world would be inadequate restitution for a disaster such as this. Finally, a civil class action is not even a remote possibility for most collective harms other than pollution, ones such as an adult's loss of political power, or a child's loss of opportunities, harms that are the cumulative result of innumerable market transactions.

The third category in my rough classification of governmental activities includes protection against harm not from others, but from ourselves. This protection, which takes such forms as drug laws and regulations, seat-belt laws, and laws against gambling, is referred to as 'paternalism'. Governmental paternalism includes restrictions upon our freedom not only for purposes of preventing us from harming ourselves, but also (less commonly) for purposes of forcing us to do good for ourselves. In short, this category of governmental activities—governmental paternalism—includes any governmental restrictions upon our freedom that are supposed to be for our own good.

Furthermore, governmental paternalism can be indirect as well as direct;

indirect paternalism being a direct governmental restriction upon others that has the effect of restricting our freedom indirectly. Examples of indirect paternalism, which is much more common than direct paternalism, include sanitation and product-safety regulations, and regulations in the form of state licensing requirements for professionals. By imposing direct restrictions upon the producers of consumer goods and professional services, these regulations indirectly restrict our freedom. In particular, they restrict our freedom to 'gamble' on consumer goods and professional services that, even though dangerous, are significantly less expensive than their regulated counterparts. Thus, by indirectly restricting our freedom to harm ourselves by purchasing dangerous goods and services, these regulations qualify as being paternalistic, indirectly paternalistic.

There may be some question whether sanitation, product-safety, and licensing regulations should, as I have been suggesting, be viewed as preventing us from harming ourselves, or whether, instead, they should be viewed as preventing us from being harmed by the producers of the goods and services in question. To the extent that, in purchasing these goods and services, consumers voluntarily assume the risk of these harms, these regulations therefore prevent us from harming ourselves. And to the extent that consumers would ordinarily not be viewed as having assumed the risk, they can nevertheless be made to assume the risk by means of appropriate written warnings, or waivers of their right to sue. Moreover, we can be sure that, in a libertarian society, the appropriate warnings, or waivers, would indeed accompany people's purchases. Law in most countries has evolved to the point where, because of alleged differences in bargaining power between consumers and producers, most such warnings, or waivers, are no longer sufficient to relieve producers from liability. But this special protection of consumers afforded by law today is fundamentally incompatible with libertarian morality, given its total commitment to freedom of contract and free-market transactions.

According to how libertarian morality is usually interpreted, libertarian rights prohibit only harm from others, not self-harm. And, of course, these rights never require self-benefit. Thus, contrary to indirect utilitarian morality, libertarian morality does not tolerate any direct paternalism, such as the prohibition of drugs, gambling, driving without seat belts, and so on. Nor, contrary to indirect utilitarian morality, does it tolerate any indirect paternalism, such as sanitation regulations, product-safety regulations, licensing regulations, and so on.

Since then libertarian morality, contrary to indirect utilitarian morality, does not tolerate any governmental paternalism, either direct or indirect, libertarian morality is therefore, to that extent, less conducive than indirect utilitarian morality to the realization of those values, such as health and safety, that governmental paternalism is designed to realize.

Of course, governmental paternalism does somewhat restrict our freedom to decide for ourselves what is in our best interests, and freedom is also an important value. Thus, along with allowing some governmental paternalism, a system of morality should, I admit, also protect us against too much governmental paternalism (see Mill 1859). Libertarian morality certainly protects us against too much paternalism by its absolute ban upon any paternalism at all. The problem is that libertarian morality goes too far; it throws out the baby with the bath water.

Indirect utilitarian morality protects us against too much paternalism as well. But the protection provided by indirect utilitarian morality takes the form of rights against the government that are much more specific than the broad libertarian rights that rule out all paternalism whatever. These indirect utilitarian rights include, as we have seen (Sect. 2.2), a right to freedom of speech, to freedom of religion, to privacy, and so on—rights that effectively disallow governmental paternalism in certain crucial areas, but stop short of an absolute ban upon any governmental paternalism whatever. Thus the protection against too much paternalism afforded by indirect utilitarian morality is considerably more fine tuned, disallowing the bad, while leaving room for the good.

The fourth category of governmental activity I want to distinguish is that of protection against, and relief from, harms that are essentially 'acts of God', rather than the consequences of voluntary human behaviour. Governmental activities protecting us against 'acts of God' include weather warnings and evacuations, emergency relief funds, containment of forest fires started by lightning, containment of beach erosion at public beaches, earthquake protection, various disease-control measures, and government-sponsored disability insurance (to the extent that it covers disabilities that were 'acts of God').

Since none of the harms relevant to category four violate anyone's libertarian rights—these rights being against only other human beings and their institutions, not against natural occurrences or 'God'—no governmental protection against these harms is allowed by libertarian morality. But it is allowed by indirect utilitarian morality. Thus, once again, indirect utilitarian morality proves more conducive to realizing, among other values, those of health, safety, and relief from suffering.

Fifth, and finally, there is the very broad category of governmental activities that are not protection against, or relief from, harm, but are the promotion of good. This category includes innumerable things such as government construction and maintenance of highways, public utilities, national, state, and city parks, playgrounds, public beaches, public museums, and all public schools, colleges, and universities. It includes government-sponsored research of all sorts, such as research into the cause and cure of diseases, and into alternative energy sources. It includes all government economic measures designed to fight unemployment and promote economic growth, such as government monetary and fiscal policies. It includes governmental measures designed to promote financial security, such as the social security programme for those over 65. In general, this category of governmental activity includes government support of many so-called public goods—goods and activities which (to the extent that they emanate from the private sector) have positive externalities (Sect. 3.1)—things such as the restoration of dilapidated houses in inner cities and vaccines against diseases. And, to the extent that poverty cannot be classified as a collective harm, this category even includes any governmental welfare measures for alleviating the suffering of the poor.

Since all libertarian rights are negative ones, they can never be violated by a mere lack of that which is good. Thus no government activities promoting that which is good can ever be for purposes of protecting us against a violation of libertarian rights; accordingly, all such activities are forbidden by libertarian morality. Once again, libertarian morality prohibits an entire category of governmental activities that indirect utilitarian morality allows. And, once again, the values that government activities in the forbidden category promote are significant: education, equal opportunity, and recreation (promoted through such measures as public schools, government aid to students, and public museums and parks); health, security, and general well-being (promoted through such measures as social security, welfare, and government-supported medical research); economic productivity (promoted through such measures as governmental monetary and fiscal policies).

In sum: of the five categories identified here, libertarian morality allows government activity only in category one, while indirect utilitarian morality allows government activity in all five. Accordingly, it would appear that libertarian morality is less conducive to the realization of those values and ideals that government activities in these other four categories promote, values and ideals that are supported by the moral intuitions of most people. This in turn suggests that libertarian morality is therefore less compatible, overall, with the moral intuitions of most people than is indirect utilitarian morality. But before we can conclude this, we must first see how the libertarian would reply to the points I have tried to make so far.

2.5 USING THE PRIVATE SECTOR INSTEAD

Of the five categories of governmental activity I have identified; libertarian morality allows category 1 and appears to exclude categories 2 through 5. I argued that, because it excludes categories 2 through 5, libertarian morality is therefore less conducive to the important values and ideals that government activities in these areas promote. The advocate of libertarian morality has two lines of defence against this argument. The first is to try to squeeze some of the more significant governmental activities that I put into categories 2 through 5 into category 1 instead, on the grounds that these activities are necessary accompaniments to the government activities of category 1—activities designed to protect us against individual harms from others. For example, the libertarian might argue that the government construction and maintenance of highways, which I put in category 5, is necessary for protecting us against a foreign attack in violation of our libertarian rights, and thus really falls into category 1. Although such an argument might work with respect to large interstate highways necessary for transporting military equipment, it probably would not work very well with respect to local, neighbourhood roads. Anyway, through enough twisting and squirming, the libertarian no doubt could indeed succeed in getting some of the government activities I put into categories 2 through 5 transferred into category 1 instead. But probably not most of these activities, and therefore I doubt if this line of defence is very promising. And, in any case, instead of all this problematic twisting and squirming, why not simply abandon libertarian morality?

The second, and I think more interesting, line of defence open to the libertarian, the one I shall concentrate on here, is to admit that libertarian morality would disallow the government activities falling into categories 2 through 5, but argue that this would make no difference because all of these activities could be performed by the private sector instead and performed more effectively. For example, people's future could be made secure through private annuities and the like, rather than through government social security. Private insurance policies could take the place of government unemployment compensation and government disability compensation. Government welfare could be replaced entirely by private charities. And so on. Furthermore—to continue with this line of defence— transferring all these activities to the private sector would promote a greater realization of what (so the libertarian would claim) is the most important value of all: freedom.

I agree that most of the governmental activities falling into categories 2 through 5 could be handled privately; a few, more effectively even. But I think that most, if handled privately, would be handled less effectively, many of them much less effectively. And some such activities could not be handled privately at all, such as, for example, protection against certain collective harms like a loss of political power, or a loss of opportunities. Finally, the net result, I submit, of transferring all these activities to the private sector would not be an increase, but a decrease in freedom. Let me try to explain.

Consider, for example, road and highway construction and maintenance. Were this done privately, the necessary funds would have to be collected, I assume, by means of tolls, which would mean stopping, again and again, to throw change into toll booths everywhere one went. The inconvenience and delays this would cause would be enormous. The general problem here seems to be an administrative one. The collection of funds necessary for constructing and maintaining roads and highways can be done, efficiently, only through a government.

Or take governmental food, drug, and product-safety regulations. These regulations are among those things that could not be duplicated privately at all.

The best that could be done privately is information on products through, for example, magazines like *Consumer Reports*, and lawsuits for any damages the products may cause (if this option were not precluded by warnings, waivers, and so on, as explained above, and if one were still alive). But private information services (supplemented by lawsuits) would not be good enough. They may have been good enough a century or so ago, but not today. Today, with modern machinery and high-speed transportation, with thousands of artificial flavours, colours, preservatives, stabilizers, pesticides, with innumerable new electrical products, children's toys, radiation-emitting products, chemicals, and drugs, today with all these new developments and more coming all the time, things are vastly different than a century ago. Today the amount of information it would take to avoid cancer, electrocution, radiation poisoning, and so on, from the innumerable products to which we would be exposed without regulations would probably fill a moderately sized library; just keeping up with the continual expansion of this information would be an everyday task.

Private certification services like, for example, the 'Good Housekeeping Seal of Approval' might help, but only up to a point. These services would necessarily leave many products unevaluated and, in addition, would create the problem of keeping track of, and evaluating, the certification services themselves, thereby creating, perhaps, more confusion than enlightenment. A proliferation of consumer publications like *Consumer Reports* might help also, but, again, only up to a point. With innumerable new products and changes in old ones confronting us continuously, either these consumer publications would leave large gaps in our information, or else they would become so numerous and detailed that most people would be unable to keep up with them. Many people would lack the time; others, the education or intelligence. Moreover, even if everyone could, somehow, keep up with all the necessary information, everyone's doing this, individually, would be extraordinarily inefficient simply because of the extensive duplication of effort this would involve. It is far better for the government, by means of regulations, to make routine, product-safety decisions for all of us than for all of us to have to try, somehow, to accumulate enough information to make these innumerable, yet routine decisions for ourselves. The amount of sheer time and effort these regulations save is enormous, not to mention the harm and suffering they prevent.

Libertarians traditionally place an exceptionally high value upon personal freedom, and, of course, governmental food, drug, and product-safety regulations do, somewhat, restrict personal freedom. By means of direct restrictions upon manufacturers and suppliers, they restrict, indirectly, our freedom to purchase whatever goods and services we may want; these restrictions are therefore indirectly paternalistic (as explained above). But the loss in freedom from these regulations is relatively minor. After all, it is hardly the freedom to decide for ourselves whether a particular pesticide or preservative causes cancer, or, given that it does, whether to eat the contaminated food anyway, that is important. What is important is the freedom to decide for ourselves matters of religion, politics, and values, freedom that, as we have seen, is protected by indirect utilitarian morality as well as by libertarian morality. Any loss of freedom that governmental food, drug, and product-safety regulations may represent is trivial indeed relative to a loss of freedom to decide these more important things for ourselves.

Furthermore, any loss in freedom that these regulations may represent is offset by a corresponding gain in freedom: namely, the freedom to purchase whatever goods and services we want without the constraints either of fear or of vast, continuous, personal investigation into their safety. Considering this significant gain in freedom, I claim that these regulations represent, on balance, not a decrease in freedom at all, but an increase. (I am presupposing here the broad sense of 'freedom' explained in Sect. 2.7 below.)

Consider, next, such things as governmental unemployment compensation, disability insurance, and social security. Although I readily admit that protection against unemployment and disability could be turned over completely to private insurance companies, even here complete privatization would reduce, not increase, effectiveness. With complete privatization, the people who needed unemployment and disability insurance the most—namely, those with marginal jobs that are often risky and always pay very little—would be the very ones who could not afford this insurance and who thus would end up with nothing. Likewise, those who earned the least during their working years, and thus had the greatest need for an annuity to support them in their retirement, would again be the very ones who could not afford one and thus, without social security, would end up with nothing also. And here I must emphasize that it is not as if most of those left unprotected could easily have been protected had they so chosen; rather it is that their poverty would make these protections available to them only at the cost of doing without basic necessities, such as adequate food for their children. Moreover, many of these people are altogether ignorant or ill-informed about these matters, thereby rendering them unable to make intelligent insurance decisions in any case. Finally, certain kinds of insurance would be much more expensive if handled privately. Take unemployment insurance. If handled privately, it would tend to be bought by those most in jeopardy of unemployment (a phenomenon known as 'selection bias'). Selection bias obviously drives the cost of such insurance way up. Only government has the ability to avoid selection bias by requiring universal coverage, thereby keeping the costs reasonable for everyone. In general, privatization in these areas would be at the expense of comprehensiveness; vast numbers of people would remain unprotected, these people being largely the very ones who needed it the most. A satisfactory degree of comprehensiveness is available only through unemployment, disability, and retirement protection that is either governmentally administered, or at least universally subsidized through generous vouchers.

And, of course, in a libertarian society there would be no governmental welfare for the vast numbers of people left unprotected to fall back upon either. According to the best estimates, the top 1 to 2 per cent of American families own about 30 per cent of the wealth in the United States today. The top 10 per cent own about 60 per cent (Avery *et al.* 1984: 865; Projector 1964: 285). And the bottom 20 per cent own virtually nothing (that is, their debts exceed, or equal, their assets). This is so even given a certain amount of governmental redistribution of wealth through progressive taxation, welfare programmes, and the like. After libertarians had done away with this governmental redistribution of wealth, and after they had done away with governmental unemployment compensation, disability insurance, and social security as well, the gap between rich and poor would be even greater; indeed, it would be enormous; people could literally be starving on the streets.

But might private charity, the libertarian remedy for poverty, possibly be an adequate remedy for all of this? To begin with, let me grant that the current system of governmental welfare in many countries is seriously flawed (see e.g. the discussion, in Sect. 5.2, of US antipoverty policy). But to rely entirely upon private charity would be no remedy, for the following reasons. First, even if private charity were sufficient in overall amount, it could hardly be as comprehensive as current systems of governmental welfare supplemented by private charity, as flawed even as most current systems are. Only the government has the wherewithall to co-ordinate aid to the poor systematically so that large numbers do not get overlooked or 'fall through the cracks'. And any lack of comprehensiveness raises a question

of justice: Is it really fair when, as between two equally deserving poor
families, one happens to receive substantial aid while, because of the
inevitable haphazardness of private charity, the other receives nothing at
all?

Moreover, would private charity indeed be sufficient in overall amount?
I doubt it. There is, of course, always the problem of people not being
generous enough. But there are two other problems with relying entirely
upon private charity that may be even more likely to cause it to be
insufficient in overall amount than lack of generosity. The first is the 'free-
rider' problem, and the second is the 'assurance' problem. The free-rider
problem arises in cases where large numbers of people reason about some
collective goal, X, as follows: enough other people will do their part for X
to be successfully realized whether I do my part or not; so I need not do my
part (my contribution would be superfluous). With large numbers of
people reasoning this way, and thus choosing to be 'free riders', obviously
X will not in fact be realized. The assurance problem, on the other hand,
arises in cases where large numbers of people reason about the collective
goal, X, as follows: not enough other people will do their part for X to be
successfully realized; so I need not do my part (my contribution would be
wasted). And, of course, if large numbers of people, through lack of
assurance, do not do their part, then, as before, X will not be realized.

Charitable giving may be subject to both the free rider and assurance
problems. Taking the collective goal of charitable giving to be the
elimination of extreme poverty (which, let us assume, everyone wants),
then large numbers of people may reason that others will contribute
enough to eliminate extreme poverty without their contribution, so their
contribution would be superfluous. Alternatively large numbers of people
may reason that others will not contribute enough to eliminate, or even
significantly reduce, extreme poverty, so their contribution would only be
wasted. The rationale here might be that the problem of extreme poverty is
simply so overwhelming that individual contributions cannot do any more
than keep some of the poor people temporarily well off enough to
encourage them to have even more children, thus perhaps making things
even worse in the long run, or at least no better. So why bother.

Or, as Allen Buchanan argues (1985: 71–3), one might not, because of
lack of assurance, fail to contribute altogether, but might instead fail to
contribute to any *large*, private charitable organization, one designed to
bring about as much co-ordinated, comprehensive poverty relief as might
be possible in the absence of any governmental programmes. The rationale
for not routing one's charitable giving through any such large organization
would be that, since the organization is private and thus incapable of
requiring sufficient contributions, chances are that its goals will not be met;
and, without its goals being met, the heavy administrative costs inevitable

with such a large organization will reduce the value of any given contribution below that of an equivalent contribution bypassing the organization and going directly to the poor family of one's choice. A widespread failure to contribute to any large, charitable organization might be almost as serious as a widespread failure to contribute anything at all because only through such large organizations can the haphazard, spotty results characteristic of private, rather than governmental, poverty relief be at least partially overcome.

And, incidentally, this assurance problem cannot be overcome simply by means of an organization's contracting to return everyone's contributions if it turned out that goals were not met. The administrative costs of returning thousands, maybe millions, of contributions, not to mention the interest on these returned contributions, would be so great as to make any such return contracts impractical and, in any case, not attractive enough to overcome the assurance problem.

Let me summarize. Perhaps neither the lack of generosity problem, the free-rider problem, nor the assurance problem would, by itself, be enough to undermine the effectiveness of relying entirely upon private charity for the relief of poverty. But all of these problems, along with the problem of a lack of co-ordination and comprehensiveness, would, taken together, be enough. Because of these problems, private charities by themselves could never be as effective as they would be if combined with well-conceived governmental programmes.

And the free-rider and assurance problems are likely to undermine not only private efforts to take over poverty relief, but private efforts to take over governmental activities of many other sorts as well. It is largely the free-rider and assurance problems that make most private efforts to protect against collective harms not just less effective than governmental efforts, but altogether hopeless. Take the destruction of the earth's ozone layer through continued use of products containing chlorofluorocarbons, products such as certain spray cans, coolants for refrigerators and air conditioners, and blowing agents for making plastic foam. Among the effects of destroying the ozone layer through continued use of these products are increased skin cancers, cataracts, weakened immune systems in humans and animals, and serious damage to plants. For the sake of protecting the ozone layer, virtually everyone no doubt would gladly give up their own use of these products in return for enough other people doing likewise. Yet if, absent any governmental prohibition of their use, enough other people will not do likewise, then one has good reason for not doing likewise oneself; one's own sacrifice would only be wasted (the assurance problem). Or if enough other people will give up their use of the products, then one also has good reason for not doing likewise; one's own sacrifice would be superfluous (the free-rider problem). So, whether enough other people will, or enough other people will not, voluntarily give up their use of the products, one has, in either case, good reason for not giving up one's own use of these products. Therefore, probably not enough people will voluntarily give up their use of these products, and (absent the availability of an inexpensive alternative to chlorofluorocarbons) the ozone layer will in fact be destroyed. In general, the free-rider and assurance problems preclude realizing many such collective goals by private efforts only; goals that include, typically, eliminating negative externalities, or achieving public goods.

The only proven way of avoiding the assurance and free-rider problems is simply by leaving people with no choice as to whether or not to do their part in realizing the collective goal. And the only proven way of leaving people with no choice is through the law backed by governmental coercion. Means for leaving people with no choice are not available in the private sector.

What we may conclude from all of this is, I think, the following. Relying upon the private sector to accomplish what governmental activities in categories 2–5 are designed to accomplish will not work. There could be no private-sector alternative at all for some governmental activities. And even in those cases where a private alternative is available, usually it will not be as effective, in many cases not even coming close. As we have seen, the reasons for this drop in effectiveness include administrative problems, the inefficiencies of duplicated effort, a lack of co-ordination and comprehensiveness, and the free-rider and assurance problems. And since the private sector is not an acceptable substitute for the governmental activities falling in categories 2–5, libertarian morality is indeed not as conducive as indirect utilitarian morality to the important values that, we have seen, these governmental activities help realize.

But, the libertarian may respond, I have not, so far, given adequate attention to what (so the libertarian would claim) are even more important values: namely, productivity and, what libertarians consider the most important value of all, freedom. Even if indirect utilitarian morality is more conducive to these other values, the libertarian may say, it is not more conducive to either productivity or freedom, and it is therefore not as adequate as libertarian morality after all.

The absolute priority that libertarians and, indeed, many other conservatives as well, seem to give to productivity and freedom is, I think, highly questionable. In my view, these values, important as they are, must nevertheless always be balanced against a number of other important values—justice, security, health, equal opportunity, and so on—and (as explained earlier) the most favourable compromise among them all must be found without giving absolute priority to any. But I shall not pursue this matter of priorities here. Instead I shall argue, in the next two sections, that even the favourite libertarian values of productivity and freedom are likely to be more fully realized with indirect utilitarian morality than with libertarian morality.

2.6 PRODUCTIVITY

Consider, first, productivity. For now, let us simply say that an economic system is productive to the extent that it produces as much as possible from a given set of resources in response to a typical pattern of demand (for a more complete explication, see Sect. 3.6).

Libertarian morality, as we have seen, entails strict, *laissez-faire* capitalism. Indirect utilitarian morality is, of course, compatible with whatever economic system is in the general welfare. We shall examine whether some version of socialism is in the general welfare in the next chapter. For now let us just assume, for purposes of argument, that indirect utilitarian morality, like libertarian morality, supports capitalism, but not *laissez-faire* capitalism; rather it supports a version of capitalism that tolerates government intervention in the economy, especially for purposes of redistributing somewhat that enormously unequal distribution of wealth which, as we saw, would result from *laissez-faire* capitalism. We may thus refer to this version of capitalism as 'welfare' capitalism. Welfare capitalism should be very familiar to us for it is, in one form or another, the version of capitalism found in every capitalistic country in the world today.

The basic libertarian objection against welfare capitalism—an objection dear to the hearts of conservatives everywhere—is that welfare undermines people's incentives to work. Supposedly, it does so in two ways. For those on welfare, it undermines their incentives to work, or work hard, simply by enabling them to get by without working hard. For the rest of the population, it undermines their incentives to work hard by necessitating such high taxes that much of their earnings for working hard will only end up going to the government in order to pay for this welfare. So, the libertarian concludes, with people's incentives to work being thus undermined, productivity naturally is less than it otherwise would be.

Now it must be admitted that there is some truth to the conservative objection to welfare. People no doubt do have their incentives to work undermined to some extent by welfare, whether they be the recipients of the welfare, or those who pay for it. But only to some extent. With welfare programmes that are properly designed, any loss of incentives will actually be very small. If, for example, the welfare takes the form of, not a generous handout to those not working, but a governmental wage supplement to those who are working, a supplement that, up to a certain point, increases as one's earned income increases, then any loss of incentives to work by the recipients of this welfare will be minimal indeed. If one chooses not to work, one does not get the supplement. An overall antipoverty policy that minimizes any recipient loss of incentives shall, in Chapter 5, be set out in detail. And if the tax scheme from which the funds for welfare come is carefully designed so that people are not faced with high marginal, or excessively high total, tax burdens, any taxpayer disincentive will be very small as well. But, once again, more on this in Chapter 5.

What I want to emphasize in this chapter is that when it comes to understanding the prerequisites for productivity, libertarians and other conservatives appear to be suffering from a hugh gap in their thinking, an enormous blind spot. Conservatives realize all too clearly that, in order for people to be productive, they must have the *incentive* to be productive. What conservatives fail to realize is that people must have the *opportunity* to be productive as well. No doubt libertarian, *laissez-faire* capitalism, with its enormous differences between rich and poor, would provide everyone with an excellent incentive ('Produce, or starve!'). But this economic system, as compared with welfare capitalism, would be woefully deficient in providing everyone with a reasonable opportunity. Wealth is opportunity. So, to be sure, for the top 1 per cent of the population, who might own perhaps 30 per cent of the country's wealth, opportunities would be excellent. But what about the vast numbers of people living in poverty? And, as we have seen, with *laissez-faire* capitalism there would probably be many more such people than with welfare capitalism. Their opportunities to do anything but the most marginally productive manual labour would, I submit, be virtually nil. Let me explain.

Take, first, education. Why, it might be asked, could not these vast numbers of poor at least become educated, and thereby develop some sort of skills more socially useful than simple, manual labour? First of all, people who grow up in the ghettos, suffering from poverty, typically never learn to appreciate the value of an education, and thus are not motivated to get one. But, secondly, even if they were motivated, in *laissez-faire* capitalism they probably would lack the opportunity anyway. Remember, all public schools and any government subsidies to aid the poor in paying for an education, even governmental loans, are strictly forbidden by libertarian morality. So all schools would be private, and probably would be very expensive indeed if the costs of private schools today are any indication. Naturally private charities, or scholarships, would help a few fortunate people here and there, but, for the majority of the poor, even elementary education would probably remain beyond their financial means. And how much opportunity to be productive does a person who can hardly read or write have?

But the explanation for why there would be few opportunities for those living in poverty under *laissez-faire* capitalism goes far deeper than just lack of education. For example, it is a fact that a child's brain development may be seriously compromised by a lack of adequate nutrition while young. Various governmental welfare programmes (food stamps, school lunch subsidies, cash benefits, and so on) in highly developed capitalistic countries have managed to keep the problem of seriously deficient nutrition under control. But, of course, all such governmental programmes would be precluded by libertarian morality. And how much opportunity to be productive does a person whose parents were not even able to fulfil the minimal nutritional needs for his or her normal brain development have?

Governmental aid in helping the poor receive adequate medical care is precluded by libertarian morality as well. But without governmental aid, the very poor, who normally cannot afford medical insurance, would be reluctant to seek any medical care except in the most extreme cases because for them it would be a very great expense indeed. And the medical care even they would seek in emergencies would often be less than the highest quality, again because of cost considerations. This understandable reluctance to seek prompt, competent, and regular medical care for themselves and their children would all too often lead to death or to a permanent disability that could have been prevented. And how much opportunity to be productive do people with inadequate medical care, perhaps even an insurmountable disability as a result, have?

Finally, capitalistic countries tend to be notoriously subject to serious unemployment, aggravated by periodic recessions. Modern capitalistic countries counter these tendencies with governmental monetary and fiscal policies, and other measures for creating jobs and stimulating growth. By careful use of these measures, most modern capitalistic countries manage, most of the time, to keep unemployment and recessions more or less under control. Other measures, such as governmentally guaranteed jobs, and perhaps even a move toward a 'share' economy as proposed by Martin Weitzman (1984), may be put to use in the future to create more job opportunities, perhaps even eliminating the unemployment problem altogether (more on these measures in Chapter 5). But all these current and potential measures for combating unemployment and recessions are, once again, ruled out by libertarian morality. The likely result, I suggest, would be unemployment vastly greater even than that found in capitalist economies today, unemployment that at times would rage altogether out of control (as it did during the Great Depression). And how much opportunity to be productive does a person for whom the economy cannot even provide a job have?

In sum, with the *laissez-faire* capitalism of libertarian morality, there no doubt would be ample incentive to be productive. But for all too many people the opportunity to be productive would be absent. All proven governmental measures for helping create the necessary opportunities are prohibited by libertarian morality. None of these measures, on the other hand, are prohibited by indirect utilitarian morality. Therefore, with indirect utilitarian morality, there exists the potential for a much more favourable balance between incentives and opportunities than there ever could be with libertarian morality. Since productivity depends largely upon there being a favourable balance between the two, I conclude that ideal utilitarian morality, and welfare capitalism, is therefore more conducive to productivity.

2.7 FREEDOM

But let us turn now to the libertarian's alleged forte: freedom. I shall try to show that, even with respect to freedom, the libertarian's most treasured value of all, libertarianism is deficient. Unfortunately, the concept of 'freedom' is notoriously controversial, and the outcome of any discussion about which, as between any given alternatives, is most conducive to freedom, depends upon which, as between alternative concepts of freedom is taken as the standard. Thus I must preface this discussion of which, as between libertarian morality and indirect utilitarian morality, is most conducive to freedom with some conceptual analysis.

What libertarians, and most conservatives everywhere, take 'freedom' to be is well known. For them, freedom is, quite simply, the absence of physical constraints, or coercion, by other people, including, of course, people in government. One is free, they say, to the extent, and only to the extent, that one's behaviour is not being physically constrained, or coerced, by other people, or by the government. If this analysis of 'freedom' were adequate, then libertarian morality, and the *laissez-faire* version of capitalism it entails, would indeed be more conducive to freedom than any alternative. This follows simply from the very broad right to freedom built into libertarian morality that forbids the government, or anyone else, from ever physically constraining, or coercing, a person for any purposes whatever—that is, beyond that of defending against, or punishing, a violation of some libertarian right.

But the libertarian analysis of 'freedom' is not adequate. A more adequate analysis, I submit, is that by Joel Feinberg in his classic essay 'The Idea of a Free Man' (1980; see also Feinberg 1973). The essence of 'freedom', according to Feinberg, is the absence of constraints upon one's ability to do, have, or be something one does, or might, desire. More specifically, statements about freedom, if spelt out fully, can usually be recast in the form indicated by the following schema: ——— is free from ——— to do (have, or be) ———. This general idea of freedom as the absence of constraints upon one would be perfectly agreeable to the libertarian if by 'constraints' Feinberg meant only ones in the form of physical constraints, or coercion, by other people. But Feinberg, correctly, recognizes that a person can be constrained in many other ways as well. He first distinguishes between positive and negative constraints. A positive constraint is the presence of something that limits one's ability to do, have, or be something. Examples of positive constraints are jails, chains, threats of punishment, and so on. A negative constraint is the absence of something that limits one's ability to do, have, or be something. Examples of negative constraints are the absence of money, the absence of strength, and the absence of knowledge. Feinberg next distinguishes between external and internal constraints. He says the simplest way to characterize this distinction is that external constraints are those that come from outside a person's body or mind, and all other constraints 'whether sore muscles, headaches, or refractory "lower" desires', are internal constraints (1980: 6). Finally, these two distinctions cut across one another, creating four categories. As Feinberg puts it:

There are *internal positive constraints* such as headaches, obsessive thoughts, and compulsive desires; *internal negative constraints* such as ignorance, weakness, and deficiencies in talent or skill; *external positive constraints* such as barred windows, locked doors, and pointed bayonets; and *external negative constraints* such as lack of money, lack of transportation, and lack of weapons. (1980: 6)

So, according to both the narrow libertarian analysis, and Feinberg's broader analysis, 'freedom' involves the absence of constraints. The big difference between the two analyses is that libertarians recognize the relevance of only one kind of constraint—external positive constraints emanating from other people or from governments—while Feinberg's analysis recognizes the relevance of any kind of constraint. I tend to think that the broader analysis defended by Feinberg is more successful in capturing what we ordinarily mean by 'freedom' than is the more narrow analysis defended by libertarians. For example, not only is it perfectly natural to say, 'Jones is not free to start taking cocaine because it is forbidden by the government' (an external positive constraint), but it is also perfectly natural to say, 'Jones is not free to start taking cocaine because her heart is not strong enough to tolerate even the smallest amount' (an internal negative constraint), or 'Jones is not free to start taking cocaine because there is no longer any available anywhere' (an external negative constraint), or 'Jones is not free to stop taking cocaine because she has become hopelessly addicted to it' (an internal positive constraint).

The libertarian will counter by claiming that the latter three constraints are not examples of a lack of 'freedom', but are instead examples of a lack of 'capacity', or a lack of 'power', and that, in general, Feinberg's so-called 'freedom' is really 'capacity' or 'power' (e.g. see Narveson 1988: 29–31).

Now I am not convinced that the libertarian is correct in claiming that freedom in the narrow sense more closely reflects ordinary language. But, for my purposes, whether the libertarian is correct does not matter much anyway. What matters more is which, as between the libertarian's narrow sense of 'freedom' and Feinberg's broad sense, represents the more important value, and therefore the value in terms of which it makes most sense to compare alternatives, such as libertarian and indirect utilitarian morality.

There can be little doubt but that the more important value, overall, is freedom in the broad sense—regardless of what one may choose to call it ('a rose is a rose by any other name'). By being free merely in the narrow sense, one enjoys the absence of only one kind of constraint—the external positive kind—whereas, by being free in the broad sense, one enjoys not only the absence of this kind of constraint, but the absence of the other three kinds as well. A constraint is, after all, still a constraint, no matter into which category it may fall. And it is precisely because freedom represents an absence of constraints upon our doing (having, or being) what we may want to do (have, or be) that we value it in the first place. Since freedom in the broad sense represents the absence of a more complete range of constraints than does freedom in the narrow sense, it therefore represents the more complete realization of that very thing for which we value freedom in the first place. Quite simply, freedom in the broad sense includes freedom in the narrow sense, and more besides.

But some libertarians support the narrow sense of freedom not because they think it more closely reflects ordinary language, and not even because they think it represents the more important value overall, but because they think that it is more subject to human control than freedom in the broad sense, and that it therefore represents the more appropriate *political* ideal (since politics has to do with what is subject to human control). The idea here seems to be that many of the constraints that limit freedom in the broad sense, such as a very low IQ, or paralysed legs, are altogether beyond human power to remedy, while the constraint of human coercion is never beyond human power to remedy (see e.g. Machan 1986: 50).

But this reason for supporting the narrow sense of freedom is weak. The line between what is, and what is not, subject to human control is not at all clear. Most of the constraints limiting freedom in the broad sense that may at first glance appear to be beyond human control are really not. It may not be within human power to raise the IQ of the person whose IQ is very low, but it is within human power to provide him with remedial training to compensate for this deficiency; and it may not be within human power to make the paralysed person walk again, but it is within human power to provide her with the latest, and most sophisticated means of artificial mobility. Of the constraints limiting freedom in the broad sense, very few indeed are not subject to alleviation through human effort. And an ideal, in order to be appropriate, need not be one the realization of which is *completely* subject to human control. For example, the mere fact that some illnesses are not subject to human control in no way compromises the appropriateness of 'perfect health for everyone' as an ideal for physicians. Likewise, freedom in the broad sense may not be completely subject to human control either, but it is certainly subject to quite enough human control for it to be perfectly suitable as a political ideal. So, in conclusion, since freedom in the broad sense is indeed suitable as a political ideal, and since it is, after all, the more inclusive and thus important value, it is therefore freedom in the broad sense, not freedom in the narrow sense, that shall serve as my standard of comparison here.

Freedom in the broad sense obviously includes innumerable different areas, or 'dimensions', in which people may be said to be free, the dimension varying according to the constraint that is in question. So in determining which alternative is most conducive to freedom overall, what we must try to do, as Feinberg says, is to calculate which alternative is most conducive to freedom in those dimensions that are most valuable, important, or significant, 'where the "value" of a dimension is determined by some independent standard' (1980: 10). Such calculations, obviously, cannot be done with mathematical precision, but are 'of necessity vague and impressionistic' (1980: 10). As vague and impressionistic as they may be, however, if they are the result of a careful consideration and balancing of the various dimensions of freedom in question, these calculations will be far from arbitrary. A careful consideration and balancing of these dimensions will reveal, I think, that, even though libertarian morality is, naturally, more conducive to freedom in the narrow sense, indirect utilitarian is more conducive to freedom in the more important, broad sense.

Let us see why. If I am right so far about welfare capitalism (which, we are assuming, is entailed by indirect utilitarian morality) resulting in more overall productivity and distributing it more evenly than libertarian, *laissez-faire* capitalism, then most people, especially the vast numbers of poor as opposed to the few rich, will be better off financially with welfare capitalism. And, of course, being better off financially, having more money, translates into less financial constraints on doing what one wants or, in other words, more of that dimension of freedom dependent upon wealth. And, with welfare capitalism, government educational and training programmes or vouchers free many from the constraints of ignorance they would be under with *laissez-faire* capitalism, thereby opening up for them innumerable opportunities. Similarly, with welfare capitalism, government health-care programmes free many from the constraints of injury and disease they would be under with *laissez-faire* capitalism.

Many other government laws and activities, permitted by indirect utilitarian morality, but not by libertarian morality, help bring about still other important dimensions of freedom. Laws such as those governing political contributions may be seen as removing certain constraints on *de facto* political power and, therefore, as contributing to political freedom. Laws creating public parks, museums, and so on, provide many people cultural and recreational opportunities to which otherwise, due to monetary and other constraints, they would never have had access. Laws prohibiting, or regulating pollution help remove insidious constraints upon enjoying our environment, and upon our good health. Laws prohibiting racial and sexual discrimination help free minorities and women from the oppressive constraints of bigotry, thus opening up many opportunities that otherwise would not have been available. And so on, and so on.

It is, in short, considerations such as these that show indirect utilitarian morality to be more conducive, overall, even to freedom than is libertarian morality—that is, to freedom in the broad sense. And, finally, if indirect utilitarian morality is thus more conducive to freedom, then this strongly suggests that it is therefore also more conducive to the closely related value of autonomy, or self-determination.

As pointed out earlier, Nozick hints that an argument in justification of libertarian morality might proceed by relating this morality to what is necessary for a meaningful life. But if, as I have argued, indirect utilitarian morality is more conducive than libertarian morality to the various values discussed above, including not only equal opportunity and productivity, but freedom and autonomy as well, then I would say that the prospects for justifying libertarian morality in terms of what is necessary for a meaningful life are dim indeed.

2.8 THE 'SEPARATE-PERSONS' OBJECTION

In reply to all of this, Nozick might well say that I have been missing the whole point of libertarianism. I have, throughout, been viewing everything just like a utilitarian does; I have been viewing everything in terms of maximizing. My arguments that indirect utilitarian morality is more conducive than libertarian morality to freedom, productivity, equal opportunity, security, and other values have all been in terms of which morality is likely to result in the greatest *overall* realization of these values. At no point have I focused on the effects upon individuals, as opposed to society as a whole. But, Nozick might say, it is precisely this society-wide focus—this utilitarian-like focus upon overall maximization—that is mistaken. The whole point of libertarianism, Nozick might conclude—a point that utilitarians seem forever unable to grasp—is simply that 'individuals are inviolable' (e.g. Nozick 1974: 30). This means that no person's well-being—that is, his or her freedom, property, life, or health—may ever, in any way, be sacrificed for the sake of another person's well-being or (what amounts to the same thing) for the sake of the *general* welfare. It is all right for people, of their own free will, to sacrifice themselves for others, but no one should ever be forced to do so.

This emphasis upon the inviolability of persons is, I think, about as deep a rationale for libertarian philosophy as libertarians have been able to come up with. It is this alleged inviolability of persons that, libertarians will say, is the main idea behind libertarian rights, rights which Nozick refers to as 'side constraints' upon the pursuit of any goals or values. And it is this alleged inviolability of persons that provides the rationale for the intense libertarian opposition to any form of utilitarianism. As Nozick puts it:

Side constraints express the inviolability of other persons. But why may not one violate persons for the greater social good? Individually, we each sometimes choose to undergo some pain or sacrifice for a greater benefit or to avoid a greater harm. . . . Why not, *similarly*, hold that some persons have to bear some costs that benefit other persons more, for the sake of the overall social good? But there is no *social entity* with a good that undergoes some sacrifice for its own good. There are only individual people, different individual people, with their own individual lives. Using one of these people for the benefit of others, uses him and benefits the others. Nothing more. . . . To use a person in this way does not sufficiently . . . take account of the fact that he is a separate person (1974: 32–3).

Nozick here is objecting to any view that calls for governments to sacrifice, in any way, the well-being of some for the sake of greater well-being of others. And this objection is directed primarily against utilitarianism—the prototype example of such a view. The general idea is that there is no social entity equivalent to separate persons that can experience the overall gain from sacrificing some people for the greater well-being of others; there are only separate persons whose well-being is thus irretrievably, and therefore wrongly, sacrificed. Let us, for convenience, call this standard objection against utilitarianism the 'separate-persons' objection. Not all utilitarian views are, of course, the same, and thus subject to the same objections. So let us see whether the indirect utilitarian morality outlined here is subject to this objection.

No doubt the libertarian will claim that the morality outlined here is indeed subject to the separate-persons objection. To be sure, the libertarian will admit, governmental pursuit of the general welfare is to be constrained, according to this indirect utilitarian morality, by negative rights. And these indirect utilitarian rights, just like libertarian rights, are to serve as 'side constraints' upon governmental activity. But, the libertarian will insist, indirect utilitarian rights differ from libertarian rights in that they are far more narrowly focused; they do not, as do libertarian rights, forbid governmental restrictions upon personal freedom in general, but merely forbid governmental restrictions of specific freedoms, such as freedom of speech, freedom of religion, and so on. Thus, contrary to libertarian rights, they leave the government ample room to pursue the general welfare. And, the libertarian will claim, any governmental pursuit of the 'general' welfare, no matter how constrained by rights, inevitably involves sacrificing, without hope of compensation, the well-being of some for the greater well-being of others. So, the libertarian will conclude, the separate-persons objection is just as applicable to the indirect utilitarian morality outlined here as it is to any other version of utilitarian morality.

I shall argue that this conclusion is mistaken; that the separate-persons objection is not applicable to indirect utilitarian morality after all. It is, of course, true that, at least within the parameters of certain constraints, even indirect utilitarian morality calls for governments to sacrifice the well-being of some for the general welfare; to act, in other words, as if society were one big person capable of experiencing everyone's experiences. And it is certainly true that there is no such 'one big person'. But it is not true that those persons whose well-being will thus be sacrificed for the general welfare have no hope of compensation. This would be true only if governmental decision-making were static; in reality, however, governmental decision-making is dynamic. Let me explain.

If the government of a political unit made only one decision throughout eternity—that is, if governmental decision-making were static—then it would be true that those chosen to have their well-being sacrificed for the general welfare would never be compensated. But, in reality, governmental decision-making is dynamic; decision after decision is made in an unending progression of decisions. Thus, although with any one governmental decision in the general welfare, a person may end up losing for the sake of others, with the next governmental decision in the general welfare, this person may end up winning at the expense of others. And if each such decision maximizes the general welfare, the odds for *any* given person are that he or she will end up better off over the long run.

Take, for purposes of illustration, a hypothetical country whose population is split evenly among those in the upper class, with 100 megaunits of well-being among them, those in the middle class, with 70 megaunits of well-being among them, and those in the lower class, with 40 megaunits of well-being among them. (A 'megaunit' is, of course, fictional, but let us suppose that well-being, or welfare, could be so measured.) Say that, during the month, this country's government has the opportunity to make the following three separate, unrelated decisions. Each concerns whether or not to create some governmental programme or other that has nothing whatever to do with enforcing libertarian rights. These decisions are separate in the sense that they cannot simply be collapsed into a single decision, as each has to be made without the government anticipating any that follow it. The first programme to be decided upon during the month increases the well-being of the middle and lower classes by 10 megaunits each, but only by decreasing the well-being of the upper class by 10 megaunits. The second increases the well-being of the upper and lower classes by 10 megaunits each, but only by decreasing the well-being of the middle class by 10 megaunits. And the third increases the well-being of the upper and middle classes by 10 megaunits, but only by decreasing the well-being of the lower class by 10 megaunits. If, as is called for by indirect utilitarian morality, the decision of each programme is favourable, the results are as in Table 2.1.

Table 2.1 Results with indirect utilitarianism

	Well-being at start of month	After programme 1	After programmes 1 & 2	After programmes 1, 2, & 3
Upper class	100	90	100	110
Middle class	70	80	70	80
Lower class	40	50	60	50

To be sure, each favourable decision sacrifices the well-being of some group for sake of the general welfare. But since a different group each time bears the sacrifice, with indirect utilitarian morality *each* group ends up better off at the end of the month. Libertarian morality, on the other hand, permits a favourable decision on none of these programmes. So libertarian morality would turn out, at the end of the month, to have been in the best interests of no group.

Obviously this illustration is, for the sake of simplicity, highly contrived. But the point it illustrates is real enough. Over the long run, any given person is, with indirect utilitarian morality, likely to be more than compensated for any temporary losses. Consequently, indirect utilitarian morality is more in the interests of anyone than is libertarian morality. With neither morality, of course, is any given person guaranteed to end up better off in the long run with it rather than with the other. But indirect utilitarian morality is nevertheless more in anyone's interests in the sense that it provides any given person with a higher *expected* utility (see Sect. 1.7 for what is meant by 'expected' utility). With indirect utilitarian morality, each person's prospects—each 'separate' person's prospects—are thus better than with libertarian morality. So, in this sense, indirect utilitarian morality, as compared with libertarian morality, is not contrary to, and thus cannot be said to sacrifice, anyone's interests. Accordingly, the separate-persons objection against indirect utilitarian morality is unavailable to those, such as Nozick, who advocate libertarian morality, or any other morality that precludes utilitarian-like decisions of the sort illustrated by Table 2.1. (That indirect utilitarianism is more in anyone's interests is defended in more general terms, and more extensively, in *Equal Consideration* (Haslett 1987*a*).)

But, the libertarian might ask, what if, in the government's quest to maximize the general welfare, each person or, as in my example, each group does not get its turn at being a 'winner'. What if, for example, legislators systematically focused only on improving the well-being of some favoured group, such as the poor? Such a group would then only receive, never give; the other groups would only give and thereby not end up better off in the long run after all.

The first thing to point out in reply to this is that, with indirect utilitarian morality, focusing only upon improving the well-being of some favoured group is strictly forbidden. All utilitarian moralities of every variety have one basic thing in common: in pursuing overall utility (overall well-being or the general welfare), the interests of everyone within the domain are to be given *equal* consideration. This means that utilitarian calculations are to give everyone's interests only so much weight as is proportional to how great these interests are, with absolutely no weight ever to be given merely for whose interests they are (see Ch. 1 for details). In view of this impartiality, it is highly doubtful that utilitarian calculations would end up always favouring some particular individual or group while always disfavouring some other individual or group. Considerations such as the diminishing marginal utility of wealth would no doubt direct a few favourable decisions towards the lower classes. But other considerations, such as the need for a certain amount of inequality for purposes of incentives, would no doubt direct a few favourable decisions towards the upper classes. In the long run, considerations such as these would tend, more or less, to balance out, sometimes favouring one group, sometimes another.

But, the libertarian may persist, surely it is improbable that, even in the long run, indirect utilitarian governmental decisions will end up favouring rich and poor more or less equally; surely if the government always did what indirect utilitarian morality prescribed, the poor would win out more often.

This may be true. But it is not significant. People are members of not just one group, but of innumerable different groups. And although, as a member of one group, a person may be able to anticipate losing through governmental decisions more often than gaining, these losses are likely to be more than compensated for through the gains made by other groups to which the person belongs. Take, for example, the rich person. He may be correct in anticipating overall losses from the government as a member of the group 'rich people'. But this person will be a member of innumerable other groups as well, such as 'automobile drivers', 'parents', 'home owners' (or even 'home owners by a beach that is eroding'), 'wild-life enthusiasts', and the group 'human beings vulnerable to diseases and injuries', a group to which we all belong. And he can correctly anticipate that, in the long run, any overall losses he may suffer as a member of the group 'rich people' will be more than compensated for by governmental decisions that, for the sake of maximizing the general welfare, benefit those in these other groups to which he belongs. Indeed, many governmental decisions, such as those resulting in food and drug regulations, consumer-protection laws, various safety regulations, environmental-protection laws, and so on, benefit almost all people, no matter what groups to which they may happen to belong.

Nevertheless, the libertarian may still persist, it is perhaps true that, if the government extracts only relatively small sacrifices from people for the sake of increasing the general welfare, everyone is likely to be better off in the long run; but what if the government extracts very large sacrifices from some? If a given sacrifice is large enough, then surely not even in the long run will the person be compensated for it. Take, for example, forcing people to be slaves, or perhaps even sacrificing their lives, for the sake of the general welfare. No amount of later gains could possibly be adequate compensation for any ultimate sacrifices such as these. Thus, the libertarian may conclude, the separate-persons objection is applicable after all.

It is here where the distinction between different versions of utilitarianism becomes crucial. For most versions of utilitarianism, this libertarian complaint may be warranted; most versions of utilitarianism are indeed committed to allowing governments to make people slaves, or even kill, for the sake of the general welfare. Although any such slavery or killings probably would not in fact be in the general welfare, nevertheless fallible governmental officials no doubt would all too often *think* that they were, and most versions of utilitarianism place no constraints at all upon their pursuit of the general welfare. But, as we have seen (Ch. 1 and Sect. 2.2), the version of indirect utilitarian methodology presupposed here is specifically designed so that, in evaluating norms of political morality, this sort of governmental fallibility is to be given full consideration. The result will be norms of political morality that place considerable constraints upon pursuit, by governmental officials, of the general welfare, constraints in the form of the indirect utilitarian rights sketched earlier, along with other norms of political morality. And, as we have seen, among these rights are not only a right to freedom of speech and to freedom of religion, but also a right not to be made a slave, and a right to life. Because of these rights, the extreme sacrifices for the sake of the general welfare, which the libertarian fears, will not be permitted by the version of utilitarian morality presupposed here any more than they are by libertarian morality.

Thus, I conclude, the claim that indirect utilitarian morality is more in anyone's best interests than libertarian morality survives these libertarian counter-arguments. There are, once again, no guarantees with either morality. But indirect utilitarian morality has, for anyone, greater *expected* utility, or well-being, than does libertarian morality (or no morality at all). Accordingly, there is still one more thing of value to most of us that libertarian morality fails to promote as well as does indirect utilitarian morality: our own self-interest.

Finally, if indirect utilitarian morality has greater expected utility, or well-being, for anyone, then it cannot be said that it decreases the well-being of some—without any hope of compensation—for the sake of the general welfare. People do have hope of compensation. Indeed, people can expect that, in the long run, they will be very much more than compensated. Therefore I conclude also that, although the separate-persons objection may be applicable to some versions of utilitarian morality, it is not applicable to the version of indirect utilitarian morality outlined here.

It is time now to look back on this chapter and see where we stand. Libertarian morality is, as we saw earlier, less conducive than indirect utilitarian morality to realizing a number of important values. These values include freedom (in the broad sense), autonomy, justice, equal opportunity, health, safety, education, productivity, and (speaking more abstractly) the general welfare. Now we see that libertarian morality is less conducive to personal well-being, or self-interest, as well. Surely all of this discredits the libertarian's claim to intuitive support. In reply the libertarian might claim that, for evaluating systems of morality, the system's conduciveness to freedom, autonomy, justice, equal opportunity, health, safety, education, productivity, the general welfare, and our personal well-being is altogether irrelevant. By then, however, the libertarian would long since have left the moral intuitions of most people far behind.

Being thus without intuitive support, the libertarian must rely entirely upon arguments. But, as pointed out earlier, *arguments* in support of libertarian morality, or the so-called inviolability of persons, are hardly the libertarian's forte. Outside of Narveson's book (1988), one finds few such arguments, much less any that are convincing. Narveson's arguments in support of libertarian morality are in terms of self-interest. As we have just seen, however, self-interest does not support libertarian morality either. One can only conclude that libertarian morality, and the *laissez-faire* version of capitalism derivable from it, are without any support at all.

3
CENTRAL-PLANNING SOCIALISM

3.1 INTRODUCTION

As I write, great changes in economic systems throughout the world are occurring. Advocates of central-planning socialism are clearly on the defensive. Yet many people still believe that current capitalism has serious moral deficiencies. Some still believe that only central-planning socialism can overcome these deficiencies and, therefore, deserves our support. To what extent are these beliefs true? From a moral perspective, how exactly do central-planning socialism and current capitalism compare with one another? These are the questions I shall address in this chapter.

As everyone well knows, however, any versions of central-planning socialism that have ever existed contained some elements of free enterprise, just as free-enterprise systems contain some elements of central planning. In short, no actual system has ever exemplified the main features of central-planning socialism fully, these features being central economic planning and public ownership of the means of production. Yet, to bring out the moral ramifications of central-planning socialism most clearly, it will be necessary to focus on a version of this system that does exemplify these features fully. Therefore, I shall not focus here on any versions of central-planning socialism that have ever existed, but on an idealized version, one that takes central planning and public ownership of the means of production to their fullest extent. This idealized version of central-planning socialism is related to actual versions of central-planning socialism exactly as the libertarian's idealized version of capitalism, considered in the last chapter, is related to actual versions of capitalism. This means that the conclusions reached here about the moral status of this idealized version are not, of course, automatically transferable to any actual versions of central-planning socialism. But these conclusions are relevant to actual versions. How relevant they are to any given actual version will depend largely upon the extent to which the actual version does, or did, in fact exemplify the main features of this idealized version.

Finally, as I did in the last chapter with *laissez-faire* capitalism, I shall proceed by comparing this idealized version of central-planning socialism with current capitalism. This procedure will generate conclusions not just about the moral status of central-planning socialism, but about that of current capitalism as well. We may take the system found in the United

States today as a prototypical example of current capitalism. But the conclusions reached here about the moral status of current capitalism are general enough to apply, more or less, to any current capitalist system.

I want to make the idealized version of central-planning socialism to be considered here as attractive a system as possible. To do so, I shall attribute two features to this system beyond public ownership of the means of production and central planning. These two additional features are a democratic political system and a particular way of distributing income that I shall explain below. Let me take each of these four features in turn and explain it in more detail.

The first feature, public ownership of the means of production (and distribution), is the one feature common to all versions of socialism. A society's means of production include both its man-made aids to production and distribution (its tools, machinery, shops, factories, and so on), and its natural resources (its farmland, forests, minerals, oil deposits, and so on). Things not properly classifiable as means of production (or distribution)—that is, things classifiable as consumer goods, such as televisions, cars, houses, and so on—are, in this system, to be privately owned.

The next main feature of this idealized system is thorough-going central economic planning. What this means is that the main economic questions of *what* is to be produced, and in what quantities, *how* is it to be produced, and *for whom* (that is, how income is to be distributed) are all to be determined, not mainly by supply and demand, as in a market economy, but by central planning. This central planning is to be carried out, largely, by public officials that make up the country's board of central planning. These central planners are to make long-term plans—say, five-year plans— that delineate how much of each good and service is to be produced during each of the five years, plans that are, of course, to be subject to revision periodically. Every productive unit throughout the economy is given a quota each year, which represents what its contribution to the overall national product that year is to be, and every unit is expected to meet or exceed its quota. And this plan will also delineate what prices goods and services are to sell for, how income is to be distributed, and how investment and growth are to proceed.

What, it might be asked, is the rationale for all this central planning anyway? Why not, as in the case of capitalism, simply allow the economy to be run 'automatically' by supply and demand? To understand the rationale, let us begin with Adam Smith's famous 'invisible hand' analogy (1776). With capitalism, Adam Smith claimed, people, in seeking their own self-interest, are led, as if by an invisible hand, to do what is in the general welfare. The idea here is simple. One seeks one's own self-interest, under capitalism, by seeking to maximize one's profits; one maximizes one's profits by maximizing sales; one maximizes sales by satisfying people's wants and desires and, finally, by satisfying people's wants and desires, one is doing what is in the general welfare.

But notice how far from complete the transformation of self-interest into the general welfare actually is under capitalism. Consider, first of all, monopolies. For those who have enough monopolistic control to do so, the best way to maximize their profits is to restrict deliberately the overall supply of what they are producing, so as to drive up its price. But deliberately created scarcities and artificially high prices are obviously not in the general welfare.

The pursuit of self-interest under capitalism also results in dangerous products and working conditions, since safety is often expensive and so would decrease profits. Moreover, capitalist owners are faced with a serious 'assurance' problem here. They have no assurance that, if they were to implement expensive safety measures, their competitors would do likewise. Without this assurance, implementing these measures might well leave them at a serious competitive disadvantage *vis-à-vis* these competitors. Thus expensive measures for assuring safe products and working conditions, no matter how worthwhile, do not get implemented.

Next, consider planned obsolescence under capitalism. The standard case is that of deliberately making products less durable than they (at little or no additional cost) could be made, so that people must therefore replace them more often than would otherwise be necessary. The more often people must replace their products, the more sales those who manufacture these products will have, and the more sales they have, the greater will be their profits. And planned obsolescence can occur in more subtle ways than this. The clothing industry, for example, does its best to change fashions each year, so as to embarrass people into buying new, more fashionable clothes even though their old clothes remain in perfectly good shape. And the electronics, automobile, and computer industries may phase in technological improvements at rates slower than necessary so as to keep continuous pressure on consumers to discard old models in favour of new and improved, current models. These tactics, although often successful in maximizing profits, are, once again, hardly in the general welfare.

Capitalism also fails to transform self-interest into the general welfare because of 'externalities'. Externalities are either positive or negative. Negative externalities are undesirable consequences of goods and services, or their production, for which those who produce or buy them do not pay. The most serious and wide-ranging negative externality is pollution. Negative externalities, being costs of goods and services for which those who produce and buy them do not have to pay, result in these goods and services being over-produced. Positive externalities, on the other hand, are desirable consequences of goods and services, or their production, for which those who produce them do not get paid. An example of a positive externality is the benefits to the general public from education, benefits such as a more productive, better informed society in which to live. Educators are paid by their students for the benefits that education confers upon these students themselves (assuming the education is private), but for the benefits that it confers upon the general public, educators do not get paid anything. Positive externalities are contrary to the general welfare to the extent that, being benefits from goods and services for which producers do not get paid, they result in these goods and services being under-produced.

Another way in which capitalism is often said to fail in transforming self-interest into the general welfare is with respect to 'public goods'. Public goods are commodities whose benefits may be provided to all people (within some given domain) at no more cost than that required to provide these benefits to one person. The classic example is national defence. If public goods were to be produced by the private sector (and they rarely are), they would have positive externalities to the extreme (since everyone fully benefits from them whether they pay or not). So absent governmental intervention, public goods, under capitalism, tend to be under-produced to the point of not being produced at all. In other words, this tendency to under-produce public goods may be viewed as, in a sense, the extreme case of the tendency to under-produce goods with positive externalities. I shall, throughout this enquiry, therefore assume that the public goods problem of capitalism is simply a special case of the general problem of under-producing goods with positive externalities.

Still another failure of capitalist economies to transform self-interest successfully into the general welfare, perhaps the most notorious failure of all, takes the form of unemployment and periodic recessions. Each individual enterprise pursuing its own best interests will lay-off workers as soon as a decrease in demand makes their continued employment no longer profitable. But their dismissal of workers will, in turn contribute to a further decrease in demand throughout the economy, which will cause still more workers to be laid off, resulting therefore in a vicious cycle that ends in a recession. Moreover, these days the damage and suffering from recessions are often made even worse through being accompanied by persistent inflation.

In short, under capitalism there are many ways in which the transformation of self-interest and private profit into the general welfare fails. And it is precisely these failures that provide the rationale for central planning. In theory, central planners have no motive other than the general welfare. So, with central planning, instead of relying upon the 'invisible hand' to somehow transform the pursuit of private profit into the general welfare (a transformation that often fails), the profit motive is bypassed, and the general welfare is pursued directly. The central planners, since they have no motive other than the general welfare, have no reason to create shortages of certain products deliberately. They have no reason for not implementing measures that will improve product safety and working conditions whenever these measures are in the general welfare. They have no reason for planned obsolescence, for not controlling negative externalities, and for under-producing goods and services that have positive externalities. And, by having control over the entire economy, they can pursue the general welfare by creating as many jobs as necessary for assuring everyone work, thereby avoiding recessions. Moreover, their control over prices implies control over inflation as well. This, in any case, is what socialist theory tells us.

But why, it might now be asked, will these central planners pursue the general welfare rather than some special interests? This brings us to the third main feature of the idealized version of central-planning socialism being outlined here: a democratic political system. According to this system, the central planners are to be elected directly by means of honest elections featuring more than one legitimate candidate for every position or, alternatively, they are to be appointed by someone who is thus directly elected and are to remain fully answerable to this person. It is this accountability to the people through democratic elections that is supposed to keep the central planners in line, to keep them from straying too far from the general welfare. (I shall have something to say about whether democratic elections are likely to succeed in doing this later.)

The fourth and final feature of this idealized version of central-planning socialism concerns how income is to be distributed. Since, with central-planning socialism, income is to be distributed not by supply and demand, but by what the central planners say, income can be distributed as evenly as might be desirable. One possibility is complete, or 'strict', equality. This alternative certainly has the virtue of simplicity, as well as administrative convenience. And it would be in accord with the very high moral value that socialists traditionally place upon equality. (An interesting case for strict equality of income, although not for socialism, is set out in Carens 1981; for a socialist case for equal income, although with qualifications, see Nielsen 1985; Norman 1982.)

Nevertheless, I doubt if strict equality would turn out to be satisfactory, even from the perspective of socialists. A strictly equal distribution of income would ignore such factors as how hard one worked, how long one worked, and even whether one worked at all, factors which, I should think, even most socialists would agree should be taken into account. Moreover, strict equality would, I suspect, undermine people's incentives to do what is socially needed. Why, after all, do any unpleasant work, or work especially hard, if one is to get the same amount regardless? Finally, complete equality does not take into account the fact that people's needs are unequal. Some, through no fault of their own, may need a lot more income than others merely to enjoy the same degree of well-being. (A case against equal income is set out by Letwin 1983: 34–45.)

Other socialists have suggested that income be distributed according to people's needs, and still others have suggested that it be distributed according to people's work effort. But there are rather obvious conceptual, not to mention practical, difficulties with these suggestions that render them unsatisfactory as well. (For a discussion of some of these difficulties, see Feinberg 1973: 107–17.)

I suggest that the best way for central planners to distribute income—the way that represents the best compromise between the socialist ideal of equality and the demands of practicality—is something like the following. A certain number of people will be needed in each occupation in order for the production quotas set by the central planners to be realized—a certain number of butchers, a certain number of bakers, a certain number of candlestick makers, and so on. The basic idea is for the income from each occupation to be set at the amount necessary for enticing just the right number of people to choose freely that occupation—the 'right' number of people in each occupation being, of course, the number needed for realizing the productivity quotas. If, say, it turns out that too many people are choosing to be butchers, and not enough are choosing to be bakers, then the income of butchers will be lowered, and that of bakers will be raised, until the right balance is achieved. But if too many people are pursuing a highly skilled profession, such as medicine, then the central planners may choose to weed out the excess not through lowering pay, but through competition instead—that is, by allowing students to continue their pursuit of the profession only if, through performance at school, scores on special tests, and so on, they have demonstrated that they are among the most qualified. The remaining students will then have to find other occupations to pursue. The point of weeding out the excess through competition rather than through income reduction is, of course, to maintain as high a quality of performance in the profession as possible. Finally, within occupations, various pay differentials might be established to reward certain kinds of socially valuable performance. All income other than that derived from officially sanctioned positions in the economy will be prohibited so as to prevent people from ignoring officially sanctioned positions for the sake of private gain on the black market, since the black market, if widespread, would undermine the central plan. (The only exception might be any gains one might happen to realize from selling one's own, bona fide, personal consumer goods that one no longer needed.)

This way of distributing income—which (as modified below) shall constitute the fourth and final main feature of the idealized version of socialism being outlined here—is attractive for several reasons. For one thing, it offers a way of getting the right amount of people in each occupation while, at the same time, preserving everyone's freedom to pursue the occupation of his or her choice. Moreover, this way of distributing income will result in a more equal distribution than is found in current capitalism. First, no one will get any income from the mere ownership of capital goods; people will get income only from labour. This eliminates one major source of the vast inequalities of income found in current capitalism. But perhaps the main source of these vast inequalities is simply this. Some people are fortunate enough to have been born with socially valuable qualities that are in scarce supply—a very high IQ, extremely good looks, an exceptionally strong and well co-ordinated body, and so on—and, with capitalism, the scarcity of these qualities enables these people to milk society for vastly higher incomes than most others will ever get. By distributing income in the way set out here, however, inequalities of income attributable just to differences in the qualities people are born with will not be as great. Let me explain.

First, notice that the enormous incomes that people with scarce, socially valuable qualities get under capitalism are by no means *necessary* for enticing these people into the occupations that put their scarce qualities to their best use—occupations such as doctor, lawyer, engineer, top business executive, professional athlete, model, actress, and so on. The greater challenge, excitement, and glamour of these occupations—that is, the greater challenge, excitement, and glamour of being a doctor, top executive, professional athlete, or actress rather than a factory worker, secretary, or shoe salesman—would be more than enough enticement for these people even without the huge incomes that these occupations command under current capitalism. Even if these huge incomes were unavailable, the president of General Motors would be in no hurry to change positions with someone working on the assembly line, and Michael Jordan, superstar basketball player, would be in no hurry to change positions with an arena janitor.

'Economic rent' is the technical name economists give any amount of income paid for a factor of production—including a scarce skill—that is above the amount that would be necessary for enticing the owner, or possessor, of this factor to put it into use. This is poor terminology, incidentally. If the technical name for this 'extra' income had to be taken from ordinary language, then something like 'economic bonus', or maybe even 'economic extortion', would be far more descriptive than 'economic rent'. In any case, the huge incomes under capitalism commanded by those in the glamour occupations are, by and large, nothing more than economic rent. The people currently in these occupations would, by and large, want to pursue these occupations even if these huge incomes were unavailable.

What the way of distributing income being set out here does is to eliminate much economic rent. To be sure, some economic rent will remain, since the income derivable from any occupation will have to equal what is necessary for enticing into that occupation the last person needed for filling that occupation's quota, thereby providing economic rent for all those prior to the last person, people who would have been willing to pursue the occupation for a lessor income. Where the saving on economic rent will be achieved is through the elimination of much competitive bidding for the services of those with scarce talents, a competitive bidding that jacks up the incomes of these people enormously. For example, in the National Football League there are a certain number of team owners, all of whom are bidding against one another (at least indirectly), and against those in other occupations, for the scarce services of the best football players. With central-planning socialism, on the other hand, the state, being the only employer in the economy, does not have to engage in competitive bidding against other employers. Thus all the economic rent that those with scarce qualities are able to extract under capitalism through this competitive bidding will be unavailable under socialism. This, combined with the elimination of income from the private ownership of capital goods, will indeed result in a more equal distribution of income than in current capitalism.

There is a flaw, however, in this method of distributing income as I have described it so far. Some people will have special needs that others do not have—such as a need for extensive medical care—special needs that could drain away their incomes, thereby turning what would otherwise have been a fairly even distribution into one that may be very uneven. These special needs are ones that do not occur regularly, as do needs for food and clothing; they instead occur only occasionally, often unpredictably, and are typically very expensive to meet. I can identify three kinds of needs in particular that are of this sort: (1) the need for medical care, including any special goods and services necessary for mitigating serious psychological or physical disadvantages; (2) the need for education, including that at the college and graduate school levels; and (3) the need for legal counsel when charged with a crime or sued. Medical, educational, or legal fees can, often unexpectedly, turn what would otherwise have been a perfectly adequate income into no more than a pittance.

To prevent this from happening, a second component of the method for distributing income that I am attributing to central-planning socialism will be this: all medical expenses, educational expenses, including those for child care, college and graduate school, and all legal expenses for those charged with a crime or being sued, are to be provided by the state to everyone at no cost. These services will not really be at no cost to anyone, of course. They will be paid for by everyone's income being just a little lower than it otherwise would be. But in this way the burden of these expenses will be spread more or less evenly throughout society rather than falling entirely upon those unfortunate enough to need expensive medical, educational, or legal services. Spreading the burden of these expenses evenly throughout society thus amounts to an important concession to the ideal of 'to each according to his or her needs', and it thereby contributes to a more even distribution of income than under current capitalism.

The third, and final component of this method for distributing income will be some sort of state-funded pension for everyone who is not working due either to disability or retirement. This pension, if at all possible, is to be generous enough to prevent any recipient from living in poverty.

This completes the outline of that idealized version of central-planning socialism to be considered here. Let us next see how, morally speaking, such a system compares with current capitalism.

3.2 COMPARING SOCIALISM WITH CAPITALISM

It might be objected that, in comparing capitalism as it is currently with an *idealized* version of central-planning socialism, I am not being quite fair to capitalism. Any currently existing system, with all its imperfections, is, it might be said, at somewhat of a disadvantage in being compared with an idealized version of some other system. Perhaps so. This comparison, however, will help bring out the major faults of current capitalism, and thus lay the groundwork for the formulation of an idealized version of capitalism that I shall set out in the remaining chapters.

There are many different criteria of morality that could be used as standards for comparing alternative systems. A socialist might suggest that such a comparison be made in terms of how well each system does in avoiding the alleged evils of competition, materialism, and individualism (cf. McLaughlin 1972).

These criteria are not, however, the most appropriate ones to use. Consider first competition. No society can aspire to much more than primitive living conditions without a strong element of economic competition. If people are not to be assigned to various occupations randomly, then, even under socialism, who gets to do what will have to be determined, at least in part, by competition. To be sure, with socialism businesses need not compete with one another as they do under capitalism; they instead are supposed to compete with themselves, so to speak, for the successful fulfillment of whatever production quota they have been assigned. But to the extent that this lack of inter-business competition may be an advantage, it is only because it eliminates the 'profit motive' of capitalism, about which I shall have more to say below.

As far as the alleged materialism of capitalism is concerned, socialist countries have always seemed just as interested in material productivity as any capitalist countries. And why not? People want and need material goods. The socialist may say that, by the materialism of capitalism, what is meant is the emphasis, in capitalism, upon the fulfillment not of genuine material needs, such as those for food, clothing, and houses, but 'false' needs, such as those for 'gas-guzzling' automobiles with ostentatious 'tail fins', and costly electronic products with so many trivial, special features built into them that the average buyer never even learns to operate them all (Marcuse 1964). And, the socialist may conclude, these false needs are created largely through the wasteful, brainwashing, propaganda from capitalist firms known as advertisements.

I am not especially impressed by this sort of argument. In the first place, I am not aware of any respectable attempts by these critics to delineate exactly how 'true' and 'false' needs are to be distinguished. And, although there are some genuine complaints to be made against advertising, it is hardly the insidious, evil thing—robbing us of our very capacity to think— that some socialists make it out to be.

On the other hand, to the extent that the complaint against capitalistic materialism is a complaint against some people waddling in obscene luxury, while others go without even the basic necessities of life, then the complaint is only too well founded. But it does not serve the cause of clarity to put forth this complaint under the heading of materialism. It is much more straightforward to put it forth instead, as I do here, under the headings of injustice and of poverty.

Finally, the complaint against the alleged individualism of capitalism is the vaguest of all. 'Individualism' can have many different senses. If individualism is used merely as the opposite of socialist collectivism, then to say that individualism is a bad thing merely begs the question against capitalism. And if the complaint against individualism is based upon a claim that the community as a whole has 'ontological' priority over the individual, then I suggest we not waste our time with such nonsense; the claim is either false, trivial, or (what is most likely) unintelligible. If, however, individualism is being used to mean excessive self-centredness and indifference towards the plight of the less fortunate, then the complaint against the individualism of those in capitalist countries is both well founded and serious indeed. But capitalist countries have no monopoly over self-centredness and indifference; these qualities have always been in abundant supply in socialist countries as well. Those socialists who take self-centredness and indifference to be a *result* of current capitalism have, I suggest, got the cart before the horse. Maybe someday people will indeed be kinder and more sympathetic towards those less fortunate, but if this day comes, I doubt if it will have been the result of a better economic system; rather, a better economic system than current capitalism will probably be the result of people having become kinder and more sympathetic.

In sum: although, among socialists, one hears much about the objectionable competitiveness, materialism, and individualism of current capitalism, I think there are more informative ways of comparing socialism with capitalism than in terms of these qualities. I propose that, instead, we compare socialism with capitalism in terms of four political ideals that are of major importance in the value schemes of most socialists and capitalists alike and, if I am not mistaken, are an important part of indirect utilitarian morality as well. These are the ideals of (1) equal access to the goods and services necessary for a decent life, (2) equal opportunity, (3) freedom, and (4) economic productivity. Exactly how to interpret each of these ideals is controversial, and I shall explain how I am interpreting them as we proceed. The first two are ideals of 'equality' or of 'justice' in that they call for an *equal* amount for everyone of something, this something being, in the one case, access to necessities and, in the other case, opportunity. The second two are ideals of 'accumulation', in that they call for *maximizing* something throughout society, with no person or group to be given any preference over any other in the process. And keep in mind throughout that none of these four ideals are meant to be absolute; all are merely prima facie. None of them, in other words, is to be realized as fully as possible no matter what the costs; all are to be weighed against other ideals, with the goal being that of realizing the most satisfactory overall compromise or trade-off. And that which constitutes the most satisfactory, overall compromise is, as I am interpreting it, that which maximizes the general welfare (see Sect. 2.2 above).

In saying that these four ideals are of major importance in the value schemes of most people, I am not, of course, claiming that these ideals are never criticized by anyone. In fact, they sometimes are criticized, especially the ideals of equal opportunity and productivity (see e.g. Young 1961). But an examination of these criticisms will reveal that they are almost always based upon taking the ideal in question to be not merely prima facie, as it should be, but absolute. Any legitimate prima-facie ideal can easily be reduced to absurdity if it is wrongfully taken as being absolute, and as thus running roughshod over all other legitimate ideals and values, no matter how dire the ramifications. Since I am not taking any ideals discussed in this book to be absolute—not even the general welfare, which is constrained by moral rights—none of these criticisms are applicable here. Occasionally equal opportunity is criticized on the grounds that, in a *laissez-faire* capitalist setting, it yields vast inequalities in wealth and income (see e.g. Singer 1979: 34–5). This criticism is, however, misplaced. The criticism should be directed instead against *laissez-faire* capitalism, not equal opportunity, since, in other settings (e.g. welfare capitalism and central-planning socialism), equal opportunity need not yield vast inequalities at all.

Finally, there are certainly a number of other legitimate ideals and values besides just the above four that would be relevant in any exhaustive comparison of alternative economic systems. Some of these other ideals and values I have touched upon already, and others I shall touch upon in what follows. But, to keep the scope of this comparison within reasonable bounds, I shall concentrate largely upon just these four ideals. Moreover, these ideals are, I submit, important enough so that any economic system that achieved the most satisfactory compromise among them would have a strong prima-facie claim to being the system that is, overall, most justified. So let us now consider capitalism and socialism with respect to each of these ideals in turn.

3.3 EQUAL ACCESS TO NECESSITIES

What, exactly, those goods and services are that are necessary for a decent life is something I shall not try to address in any detail here. The goods and services necessary for a decent life do, I suppose, vary to some extent from society to society, but only to some extent. Common to people everywhere is the need for adequate food, clothing, shelter, protection from crime, adequate medical care, and a healthy environment. Moreover, in any large industrial society today, necessary goods and services include, I would say, education beyond even high school for those both interested and qualified, and legal services if accused of a crime or sued. And surely a decent life includes a certain amount of entertainment and leisure as well.

There are many different ways in which people might have 'equal' access to these things (we may refer to them simply as 'necessities'). One way is for there to be a decent job available for everyone able to work—a job that paid enough to purchase these necessities—and adequate welfare for everyone unable to work. Thus, as I am interpreting it, equal *access* to the goods and services necessary for a decent life need not necessarily mean equal *possession* of these goods and services; a person may have access to these goods and services without possessing them by, say, choosing not to work even though a decent job is available, or even by choosing not to purchase them even though the person's income is sufficient for doing so.

In realizing this ideal of equal access to necessities, the socialist system outlined here appears to have at least three significant advantages over current capitalist systems. First, this system provides full employment or, in other words, a decent job for everyone able and willing to work. With central planning, as many jobs as necessary for eliminating unemployment can be created. This contrasts with most capitalist systems today in which an unemployment rate somewhat over 5 per cent has come to be viewed as 'natural', and in which, during periods of recession, unemployment can exceed 10 per cent. Such unemployment generates much poverty and suffering.

Secondly, as already explained (Sect. 3.1), the socialist system outlined here has the advantage of distributing income more equally than it is distributed in capitalist systems today. So, other things being equal, the lowest paying jobs in this socialist system will tend to pay more than under current capitalism, and thus, other things being equal, working people's access to necessities will tend to be greater.

Thirdly, with this system the educational, medical, and legal services necessary for a decent life are to be borne by the state. By thus spreading the burden of these expenses more or less evenly throughout society as a whole, no one is therefore denied access to at least these vital services. There is no reason, of course, why the expenses for these services could not be borne by the state in capitalist countries as well, and in some capitalist countries today, such as Sweden and Great Britain, these expenses are, in fact, borne largely by the state. But, to the extent that a capitalist system does not provide these benefits, then, to that extent, the socialist system outlined here will, once again, be more successful in realizing the ideal of equal access to necessities.

3.4 EQUAL OPPORTUNITY

Next, consider the ideal of equal opportunity. Just as the essence of 'freedom' is the absence of *constraints* upon doing, having, or being something (Sect. 2.7) the essence of 'equal opportunity' is the absence of *disadvantages* in competing for something. More specifically, statements alleging the existence of some equal opportunity, if spelt out fully, can usually be recast in the form indicated by the following schema: ——— (fill in a person(s) or group(s)), compared with ——— (fill in another person(s) or group(s)), does not suffer from (fill in a disadvantage(s)), in competing for ——— (fill in an object(s) of competition). The 'objects of competition' with respect to that equal opportunity being taken as a political ideal here are (desirable) social positions or occupations—doctor, entrepreneur, public official, football player, and so on. The people in question are, of course, everyone (in the political community) compared with everyone else. And the disadvantages that preclude equal opportunity for everyone in competing for these positions include any disadvantages whatever that, if everyone had an equal desire to compete successfully, would generate a higher probability of some competing more successfully than others.

Full realization of equal opportunity of this sort for all would mean that, in competing for, and within, the various social positions available in society, no one suffered from any mental, physical, financial, or social disadvantages at all. So, obviously, its full realization is impossible. And even if equal opportunity could be realized fully, it would be at too great a sacrifice in the realization of other ideals for its full realization to be justifiable.

Nevertheless, most of the disadvantages that are the source of unequal opportunity can, if society so chooses, at least be mitigated, and at not too great a sacrifice. Mental disadvantages can be mitigated through education, counselling, and treatment for mental illnesses or personality disorders. Physical disadvantages can be mitigated through decent medical care available to all, proper equipment for alleviating physical handicaps, and so on. Financial disadvantages can be mitigated through a more equal distribution of income and wealth. Finally, social disadvantages can be mitigated through measures that combat racial, sexual, and other forms of unjustifiable discrimination.

In mitigating the disadvantages that create unequal opportunities, the socialist system outlined here will be more successful than current capitalism. I do not think socialism will necessarily be more successful in mitigating unjustifiable discrimination, except to the extent that it does so indirectly by alleviating poverty and vast differences in wealth. But this system does have features built into it that will, I think, mitigate mental, physical, and financial disadvantages more successfully than current capitalism.

Consider first how, with this socialist system, all educational expenses, including college and child-care expenses, and all medical expenses, including even those for alleviating any serious mental or physical handicaps, are to be paid by the state. Few things cause opportunities to be more unequal than do inequalities in education and medical care. And few things cause medical care and education to be more unequal than having the expenses for them borne entirely by individuals rather than by the state. Many individuals simply will not, or (as is more frequently the case) cannot bear these expenses themselves. Thus, unless these expenses are borne by the state, these individuals, and their children, will end up with serious mental and physical disadvantages relative to those who do get adequate education and medical care. And, of course, these disadvantages translate into unequal opportunities.

Now, to be sure, in every capitalist country today some educational and medical expenses are borne by the state, just as they are in the socialist system outlined here. In Great Britain, for example, medical expenses are borne by the state, and so, supposedly, are the expenses for higher education. But both British programmes have drawbacks, especially the one for higher education. The main drawback of the latter programme is that it limits the amount of children who eventually attend college or a 'polytechnic' school in Britain to, merely, about 15 per cent; clearly British children are not so stupid that only 15 per cent would benefit from higher education. In the United States, by contrast, nearly 50 per cent of all children attend college. Yet since the burden of college expenses in the United States is not borne by the state, many children cannot, for financial reasons, attend a college compatible with their capacities, and many others cannot attend any at all. To the extent that the programmes currently in force in capitalist countries today have serious limitations, such as the British programme for higher education, or to the extent that there are no government programmes at all, the socialist system outlined here provides more equal opportunity.

But even if medical care, and education from child care through college, were equally accessible to all, the rich child, merely as a result of his or her family's wealth, is still likely to have numerous small advantages over the poor child, all of which add up to a very big advantage. The rich child is likely to have a private bedroom in which he or she can read, think, or study in peace and quiet, but not the poor child. The rich child is likely to have books, and other means of intellectual stimulation readily available, but not the poor child. The rich child is likely to have three nutritious meals every day, but not the poor child. The rich child is likely always to be well clothed, and well equipped with stimulating toys, sporting equipment, and so on, but not the poor child. The rich child may be given private lessons in music, dancing, sports, and many other things, but not the poor child. And so on, and so on.

These disadvantages from which a poor child suffers cannot be eliminated altogether. But, to the extent that the gap between rich and poor is less great to begin with, these disadvantages will be mitigated, and opportunities will thus be more equal. Since, as we have already seen, the gap between rich and poor will indeed be less in socialism than in current capitalism, this is another reason why socialism will be more successful in realizing the ideal of equal opportunity for all.

One ramification of the more equal distribution of wealth under socialism deserves special mention. Wealth provides the rich with political opportunities not available to the poor. It is a well-documented fact that, in a democracy, the greater one's wealth, the greater one's opportunity for occupying either the position of governmental official oneself, or that of 'kingmaker', or even that of being 'the power behind the throne'. In 1982, for example, 82 per cent of Senate elections in the United States were won by the candidate who spent the most on his campaign (Cohen and Rogers 1983). One reason why wealth opens up political opportunities is that wealth gives one an advantage in influencing public opinion. If freedom of speech is recognized and protected, then, of course, the opportunity to influence public opinion is, in theory, available to everyone. But to present one's ideas to others in such a way as to have real influence upon public opinion generally requires money—money to buy time on radio and television, to buy space in newspapers and magazines, or to publish books and pamphlets. Rich people have this money, poor people do not. So the rich, but not the poor, can, and do influence who is elected simply through large campaign contributions. Moreover, after these officials are elected, the rich, but not the poor, can, and do influence their decisions through expensive lobbying. In short: although everyone in a democracy may have the same *de jure*, or formal, political opportunities, people's *de facto*, or actual, political opportunities vary greatly with their wealth. It follows that, other things being equal, the more equally wealth is distributed throughout society, the more equal will be people's political opportunities. In the socialist system outlined here, wealth will be distributed more equally than in current capitalist systems; therefore, other things being equal, so will political opportunities, which is indeed an important ramification. (But bear in mind that equal opportunities are of little value if they mean equal lack of opportunities; on this, see the next section.)

Finally, socialism will be more successful in realizing the ideal of equal opportunity simply because it eliminates all means of making money merely from money. Let me explain. Among the social positions of value to which capitalism gives rise is that of (private) investor. Pure investors merely invest in the ventures of others, by purchasing stocks, bonds, and so on, or perhaps by renting productive goods they own to others. Entrepreneurial investors, on the other hand, use their wealth (and lines of credit) to buy the necessary goods and services for carrying out their own business ventures. To the extent, however, that entrepreneurs use the wealth of others to finance their business ventures, they are not investors, but managers of other people's wealth. Entrepreneurs are typically both investors and managers. But it is not upon the position of entrepreneur, but of (private) investor that I wish to focus here. The opportunity to occupy this position varies proportionally with one's wealth, obviously. And since wealth is distributed very unequally in the typical capitalist country today, so is the opportunity to be an investor. In the typical capitalist country there will be many people with no investments at all, these people corresponding, for the most part, to those with virtually no wealth at all. There will also be many small-time investors, these corresponding, for the most part, to those with only modest wealth. These will be the people with a modest stake in some pension fund that invests in stock, or with a few token shares of stock in the company for which they work. From these investments, no one will get rich. Finally, there will be a relatively few really big-time investors—people with lots of wealth to 'play' with. In the United States today, big-time investors will be found only among the richest 10 per cent, or so, of the population, which owns about 60 per cent of the wealth; most will be found in the top 1 or 2 per cent, which owns about 30 per cent of the wealth (Avery *et al.* 1984: 865; Projector 1964: 285). According to one study, the richest 1 per cent owns approximately 60 per cent of all wealth in the form of privately held corporate stock (Greenwood 1973: 35). These are the people whose investments do earn huge sums of money. It is thus no myth that, under capitalism, the rich get richer, and the poor get poorer. The rich invest their money to make still more money; the poor must use all their money just to say alive, leaving them therefore with nothing.

I am not claiming, as socialists often do (e.g. Schweickart 1980: ch. 1), that big-time investors are mere parasites of society. Some, who do nothing in life other than turn their inherited fortunes over to investment managers no doubt are parasites. But others, managing their own investments, may work hard and, at the very least, perform the socially useful function of determining, after carefully reviewing the alternatives, which ventures are worthy of support and which not. What I am claiming, however, is that the opportunity to be a big-time investor is extremely unequal. Only the rich have this opportunity, which serves to make all the more significant the fact that, under current capitalism, opportunities to become rich in the first place are so unequal.

With socialism, on the other hand, personal investment opportunities are not unequal; no one has any such opportunities at all. Any investments are made by the central planners in the name of the people as a whole. Since the social position of private investor does not even exist, personal investment opportunities are obviously no obstacle to equal opportunity or a reasonably equal distribution of wealth with socialism. With socialism, in other words, mere money cannot make money, only honest labour can.

For the various reasons mentioned above we must conclude, I think, that the system of central-planning socialism outlined here will be more successful, overall, than any current capitalist system in realizing the ideal of equal opportunity.

3.5 FREEDOM

The third ideal in terms of which I shall compare socialism with capitalism is freedom—but not freedom in that narrow sense espoused by libertarians and other conservatives (a sense that, it so happens, helps direct attention away from the grave moral deficiencies of their position). Freedom in this narrow, libertarian sense is simply the absence of coercion by others, including the government. The ideal of freedom in terms of which I shall compare these systems is, instead, freedom in that broad sense explained in Chapter 2. In this sense, freedom not only can be the absence of coercion—an 'external positive' constraint—but also can be the absence of any of the three other kinds of constraints identified by Joel Feinberg (1980; see also Feinberg 1973). According to this broad sense of freedom, a black person, for example, may lack the freedom to pursue a career as a brain surgeon not only through a governmental prohibition upon blacks pursuing this career (an external positive constraint), but also through racial discrimination so widespread that, even though it would be perfectly legal to do so, no medical schools ever accept blacks (an external negative constraint), through not having the IQ to master the necessary skills (an internal negative constraint) or through having been hopelessly conditioned by the racially biased society into believing this career to be unsuitable for black people (an internal positive constraint). Those who espouse freedom in the narrow sense consider only external positive constraints to be relevant. But freedom in the broad sense, where the relevant constraints include not only external positive ones but the other three kinds as well, is the more appropriate ideal in terms of which to compare alternative systems. It is more appropriate not necessarily because it is a more accurate representation of what, by 'freedom', we mean in ordinary usage (although I think in fact it is). Rather, it is more appropriate because it is the more inclusive ideal and, as such, the more significant, or important, of the two (as explained in Sect. 2.7).

Remember that the ideal of freedom being used here is put forth as an accumulative, not equalitative, ideal, which means that what counts as a more complete realization of this ideal is a greater total amount of it throughout society, not, as in the case of the first two ideals, a more equal distribution of it. And, as explained earlier (Sect. 2.7), since people can be free in the broad sense through the absence of constraints falling into any one of the four categories referred to above, people can therefore be free in the broad sense in many different ways; freedom, in other words, has many different 'dimensions', some of which are more important than others. Moreover, 'constraint' is not an all-or-nothing concept. There are degrees of constraint; accordingly, there are degrees of freedom. So determining whether one economic system gives rise to a greater total amount of freedom throughout society than another involves determining the degree of freedom realized by each system along different dimensions, weighing the importance of these dimensions against one another, and thereby seeing which system provides the most favourable, overall 'combination' of freedoms. In principle, what counts at the most favourable, overall combination of freedoms is, I submit, that combination most in the 'general welfare' (see Sect. 2.2 and Ch. 1). So, since the question, 'which alternative combination of freedoms is most in the general welfare?', always has a correct answer (Ch. 1), the question, 'which system provides the greater total amount of freedom?', always has, I submit, a correct answer as well. That is, it always has a correct answer in principle, although, in practice, any conclusions about what the answer is must remain impressionistic and controversial. Yet, if supported by careful argumentation, any such conclusions will certainly not be arbitrary.

Perhaps the most general way in which people are constrained from doing, having, or being what they may desire—that is, the most general (though not necessarily most important) way in which people lack freedom—is through having insufficient wealth. We may refer to this dimension of freedom as 'financial' freedom. Using Feinberg's terminology, insufficient wealth is an external negative constraint. The constraint of insufficient wealth can be mitigated in two different, sometimes incompatible ways. First, this constraint can be mitigated by increasing the overall, total amount of wealth there is in society—that is, by increasing productivity. Secondly, this constraint can be mitigated by distributing the amount of wealth there is in society more evenly.

Why, it might be asked, does a more even distribution of wealth result in more overall freedom? Let me try to explain. Consider first the diminishing marginal 'desire satisfaction' of wealth (that is, its diminishing marginal capacity for satisfying desires). This should not be confused with the well-known diminishing marginal *utility* of wealth. Most economists and many philosophers make no distinction between desire (or preference) satisfaction and utility. For them, therefore, diminishing marginal desire satisfaction and diminishing marginal utility are one and the same thing. I argued in Chapter 1, however, that desire (or preference) satisfaction and utility are not quite the same thing. Thus neither are the diminishing marginal desire satisfaction of wealth and the diminishing marginal utility of wealth, although they are indeed closely related. And, as general tendencies (not ironclad rules), they are both equally sound.

The diminishing marginal desire satisfaction of wealth means that, generally speaking, as one's wealth increases, the personal desires that each additional increment of wealth goes to satisfy will become increasingly less urgent or important. The undeniable common sense behind this is simply that, with one's first, say, thousand dollars one will, generally, satisfy what one considers one's most important desires, with one's next thousand dollars, one's next more important desires, and so on, with the desires that each additional thousand dollars goes to satisfy generally becoming less important than those satisfied by the previous thousand dollars. For example, if John Filthyrich were to find an unclaimed thousand dollars in the woods at a time during whch he was still penniless, then he would probably use the money to satisfy his desires to eat and stay alive. If, however, he found another thousand dollars after he had become a multimillionaire, then he might, say, use the money merely to buy an extra piece of decorative trim for his latest yacht. Satisfying his desires to eat and stay alive would have been more important to him than satisfying his desire for an extra piece of decorative trim.

Now I happen, personally, to believe that the diminishing marginal desire satisfaction of wealth tends to be applicable not just *intra*personally as we have just seen, but *inter*personally as well. For example, if, say, the government were (through taxes) to take a thousand dollars away from John Filthyrich, a multimillionaire, and give it to Joe Pauper, who is starving on the streets, then the desires that this thousand dollars would go to satisfy—Pauper's desires to eat and stay alive—would, I believe, probably be more important than the desires that this thousand dollars would have satisfied had this money been spent by Filthyrich on himself instead.

But the argument here does not depend upon the diminishing marginal desire satisfaction of wealth being applicable interpersonally. Consider, first, the case of utility. When combined with the fact that we do not yet know how to make valid interpersonal comparisons, it follows merely from the *intra*personal diminishing marginal utility of wealth that the more evenly a nation's wealth is distributed, the more utility, overall, is likely to result from this wealth. Although all of this makes perfectly good sense without the help of mathematics, it has also been proven mathematically (Lerner 1946: 28–31; see also Buchanan 1985: 56–8; Brandt 1979: 311–16). Likewise, from the (intrapersonal) diminishing marginal *desire satisfaction* of wealth, it follows that the more evenly wealth is distributed, the more overall desire satisfaction there is likely to be (taking into account both the number and importance of the desires satisfied).

Finally, if the more evenly wealth is distributed, the more overall desire satisfaction there is likely to be, this means that, the more evenly wealth is distributed, the less financial constraints, overall, there are likely to be upon the satisfaction of desires. But the less financial constraints, overall, there are upon the satisfaction of people's desires, the more overall financial freedom there will be throughout society (since financial freedom just is the absence of financial constraints upon the satisfaction of people's desires). And, of course, the more financial freedom there is throughout society, then, other things being equal, the more freedom in general there will be. So we may conclude, finally, that the more evenly wealth is distributed, the more overall freedom there will likely be—other things being equal, of course.

But we have already seen that central-planning socialism will distribute wealth more evenly. Therefore, to that extent at least, it will likely result in more freedom overall throughout society than current capitalism. Of course this gain can be more than cancelled out if, with socialism, the overall, total amount of wealth throughout society—i.e. productivity—will decrease. The complex topic of the effect that central-planning socialism will have upon productivity is investigated in the next section.

A more equal distribution of wealth has a subtle effect upon still another dimension of freedom. One's wealth typically determines, to a significant extent, how much respect one gets from others. And how much respect one gets from others typically determines how much *self*-respect one has. Moreover, it is clear that a lack of respect from others, and a lack of self-respect, can, in innumerable different ways, constitute powerful constraints upon one's being able to do, have, or be what one may desire—a lack of respect from others being an external negative constraint, and a lack of self-respect being an internal negative constraint. These constraints are often called 'class' barriers. They can, for example, preclude one from being friends with, or dating and marrying, certain people, from attempting to pursue certain careers, and so on. Let us, for convenience, refer to the dimension of freedom in question here as 'social' freedom, and the corresponding constraints as 'social' constraints. A more equal distribution of wealth will not only mitigate financial constraints, thereby increasing financial freedom, but it will also mitigate social constraints, thereby increasing social freedom.

Next, to the extent that people must pay for their own educational, medical, and legal expenses, many of the constraints of ignorance, disease, and injustice can be avoided only by the reasonably well-off. To the extent, however, that the state pays these expenses instead, everyone, not just the reasonably well-off, can avoid many of these serious constraints upon freedom. To be sure, the moderately lower incomes that people must tolerate in order to support these state benefits will mitigate somewhat the freedom that these benefits provide. But—assuming only . that these benefits do not significantly affect overall productivity (and, as I argue in Chapter 5, they need not do so)—any such loss in freedom from these benefits will be far outweighed by the gains in freedom. This follows largely from, once again, the diminishing marginal desire satisfaction of wealth, along with how extremely serious the constraints are that these benefits serve to remove. Educational, medical, and legal expenses are, as we have seen, paid by the state in the socialist system being considered here. So, to the extent that these expenses are not paid by the state in capitalist systems today, this socialist system will be more successful in furthering overall freedom in these ways as well.

But let us now turn to some dimensions of freedom that seem to be the most successfully promoted in current capitalist systems. First, take entrepreneurial freedom. This is the freedom to start, run, and reap the profits from whatever business or productive enterprise one wants. It is true that, even with the extreme version of central planning being considered here, there is freedom of occupation at least in the sense that you can compete freely for whatever occupations are available and selections are to be made only on the basis of people's genuine qualifications. But the occupations that are available are limited only to those that the central planners make available. This means that you cannot freely pursue your own idea for a business—not, at least, unless your idea is already part of the central plan or you can convince the planners to make it a part of the plan. And if they did approve your idea, then, even if you were put in charge, the business would have to be run more or less according to the planners' general specifications, not yours, and you would not reap the profits from the business yourself, but would merely receive a salary like anyone else. To be sure, your salary would likely be somewhat higher than the salaries of those under your supervision (even in socialism, some differentiation in salaries will be desirable for the sake of incentives). But your salary would not compare very favourably with the profits taken in by many capitalist entrepreneurs.

In response, the advocate of socialism may belittle the importance of entrepreneurial freedom, pointing out that it is essentially a freedom enjoyed under capitalism only by the rich, those with enough wealth to become entrepreneurs. But this response would not be entirely appropriate. It is certainly true that to be a big-time investor one must be rich (see Sect. 3.4 above). And being rich certainly helps in being an entrepreneur. But entrepreneurs need not necessarily use their own wealth (or lines of credit) for financing their ventures; there are, under capitalism, ways of raising money from others for promising ventures. These ways include the sale of bonds, or shares in one's venture, or soliciting the support of someone who is rich. So entrepreneurial freedom is not enjoyed only, even if primarily, by the rich. Moreover, for socialism to preclude any private entrepreneurial endeavours is, I suggest, to preclude an important outlet for personal creativity—creativity that, under central-planning socialism, can be exercised almost only by the central planners. In sum, the dimension of freedom that entrepreneurial activity represents is not, I think, insignificant. It also represents an important source of productivity, but more on this later.

Next, capitalism succeeds better than central-planning socialism in providing consumer freedom. By 'consumer' freedom, I mean the freedom to choose what goods and services the economy is to make available. As is well known, in capitalism the goods and services that are made available are those that are 'demanded'; those that have been chosen by people's dollar (pound, franc, yen, or whatever) 'votes'. For example, if people generally prefer football games to operas, thereby spending their money on football games, not operas, then football players generally will make more than opera singers, more people will pursue football rather than opera as a career, football stadiums rather than opera houses will be built, and so on. In other words, capitalism automatically accommodates itself to people's wants and desires. With central-planning socialism, on the other hand, people must accept only those goods and services that the central planners choose to make available. The central planners will, no doubt, try to accommodate consumer preferences to a large extent. But they will not be as successful in doing this as is a capitalist economy, and for these reasons.

First, the central planners, having the awesome power of deciding what goods and services are to be produced, will no doubt at times choose, for the sake of what they consider to be a greater good, to exercise this power contrary to people's preferences. They will, in other words, act paternalistically. If, for example, the central planners believe that operas are better for people than football games, then, contrary to people's preferences, the central planners may choose to have opera houses rather than football stadiums built. Since I have stipulated that the central-planning system under consideration here is a democracy, people, if dissatisfied enough with the planners' choices, could, of course, always refuse to re-elect them. But this is not an entirely adequate remedy; many opera houses can be built in the four or five years between elections. And to have elections much more frequently than this would not allow for sufficient continuity. Moreover, majority decisions do not reflect the intensity of people's preferences, as does the market.

To be sure, people's preferences are not, as I argued earlier (Ch. 1), definitive of what is in their best interests. Therefore, as I also argued (Sects. 2.4 and 2.5), a certain amount of governmental paternalism—especially indirect paternalism—is not only desirable, but will actually increase overall freedom. So my point is not that governmental paternalism is never justified. My point is simply that, with central-planning socialism, governmental paternalism will be so easy that central planners will, almost inevitably, go beyond the limitations that are appropriate.

And exactly what limitations upon paternalism are indeed appropriate? Take, in particular, governmental paternalism in the form of protecting us from harmful products, both goods and services. Although I do not want to pursue this complex question in much detail here, what I have in mind are limitations (or general guidelines) something like the following. First and foremost, governmental paternalism should be limited to cases in which (a) the alleged harm from the product is something that virtually everyone would readily agree is in fact harmful—such as cancer, food poisoning, serious injury, and so on—and (b) the connection between the product and this harm is well established scientifically. This guideline rules out such objectionable forms of paternalism as, for example, restrictions upon freedom of speech or religion. I would suggest also that governmental paternalism be limited to cases in which one or more of the following additional conditions are met: (1) to many consumers, the connection between the product and the harm is not readily apparent; (2) the product is so addictive that many consumers are unable to discontinue its use when they want to do so; (3) safe, similarly priced products that are (as virtually everyone would agree) perfectly adequate substitutes for the harmful product are or can be made available; or (4) the product can, without an enormous increase in its price, be changed in such a way to make it far safer.

Maintaining paternalism within something like these guidelines is certainly not out of the question for a capitalist government in which proposed paternalistic regulations will typically be fought tooth and nail by a powerful, independent business community. But with central-planning socialism, because of the central planners' extraordinary control over the entire economy, maintaining paternalism within something like these guidelines will be unlikely.

The second, and even more important, reason why central-planning socialism will not be as successful as capitalism in providing consumer freedom is this: central planners can never have enough knowledge of consumer preferences to be as responsive to them as capitalist economies are through supply and demand. They can never be as responsive even if (as will be unlikely) they want to be. More on this later.

Furthermore, in coming up with new and better ways of meeting consumer preferences, central planners cannot be expected to match the collective innovativeness of the countless capitalist entrepreneurs who, with their very economic existence depending upon it, are each seeking new and better ways.

Or we might look at it this way. No one—certainly not central planners—can adequately predict the relatively few innovations that will really be successful. But only capitalism allows for the free-market competition among all innovations that is necessary for these few successes to emerge. To put it metaphorically, only capitalism allows the many seeds to be planted that are necessary for the few to bloom. Central-planning socialism has no way to match the fruitful trial and error that this free-market competition provides.

One ramification of capitalism's innovativeness will be more new products, and more meaningful variations of the same kind of product, than can ever be generated by socialism. So not only does a capitalist economy provide more consumer freedom by being more responsive to consumer preferences, but it also does so by providing a wider range of goods and services from which consumers may choose.

Socialists may reply by arguing that entrepreneurs under capitalism do not so much respond to consumer preferences as create them, through advertising. And, among other things, this leads to what economists call 'monopolistic competition', the ability of each competitor, due to minor differences in its product, to raise its prices artificially. There is no doubt some merit in such a response, but not much. Socialists tend, I think, to exaggerate the ability of advertising to influence consumer preferences through manipulating people to behave irrationally. Indeed, some socialists carry on about this almost to the point of paranoia (e.g. Marcuse 1964). People cannot, however, be manipulated through advertisements quite as easily as many socialists think. Otherwise, all any large corporation would need for successfully marketing a new product is a large advertising budget but, as the 'Edsels' of this world prove, success does not come quite so easily. And to the extent that advertisements do succeed in creating preferences, they often do so not by manipulating us to behave irrationally, but by providing information that appeals to our rationality. And to the extent that advertisements influence us by means of rational appeals, they do not decrease consumer freedom, but increase it by helping somewhat to alleviate the constraints of ignorance in making consumer choices. To be sure, much advertising today is more inane than informative, but inane advertising is objectionable not so much because it succeeds in 'brainwashing' us, but because it is a waste of money, a waste of the necessarily limited resources at a society's disposal.

A much more telling response to the argument that capitalism provides more consumer freedom than socialism is this. Capitalism is supposed to provide more consumer freedom, since the goods and services the system makes available are determined by the dollar 'votes' of consumers themselves, rather than by central planners. The catch is that these so-called votes are distributed unequally, the rich having vastly more of them than the poor. The inevitable result, of course, is that the system thus turns out yachts and diamond necklaces in response to the vast number of votes by the rich before turning out even basic necessities for the poor, who have little or no votes to which the system can respond. Is this then a reflection of people's preferences in general, as opposed to merely the preferences of the rich? Hardly. A referendum in which some have vastly more votes than others does not reflect people's preferences accurately. Capitalism, it will therefore be concluded, provides consumer freedom only for the rich. This, the critics of capitalism will claim, leaves the argument that capitalism provides more consumer freedom than socialism very much in doubt.

In reply, it may be said that some people have freely chosen not to make the sacrifices in training, hard work, and so on, necessary for having as many dollar votes as the rich, and therefore they have no legitimate complaint. True enough.

In typical capitalist systems today, however, many others lack sufficient dollar votes even for basic necessities, not by choice, but by discrimination, by a lack of inherent capacities, or by other misfortunes altogether beyond their control. And all the while, the rich bask in their luxuries. Any inequality of dollar votes that is this extreme deserves to be labelled 'excessive'. And, just as the critics of capitalism claim, excessive inequality does indeed cast doubt upon the argument that capitalism provides more consumer freedom than central-planning socialism.

But, I suggest, it casts doubt upon the argument only with respect to those capitalist systems in which the inequality in dollar votes is, in fact, excessive. In those capitalist systems in which the inequality is not excessive—and by 'not excessive' I mean virtually everyone has access to dollar votes sufficient at least for basic necessities—consumer freedom, although not unlimited, will still, I suggest, far exceed that found in central-planning socialism. Moreover, an excessively unequal distribution of dollar votes, or wealth, and any corresponding serious lack of consumer freedom, is far from a necessary feature of capitalism (as I try to show in the next three chapters), whereas serious lack of consumer freedom is, I submit, a necessary feature of central-planning socialism.

Next, as Milton Friedman has argued (1962: ch. 1), still another dimension of freedom that capitalist systems realize to a greater extent than centrally planned socialist systems is freedom of speech and, more generally, political freedom. The argument goes like this. In order to have effective political freedom, people must feel free to express their political opinions no matter how critical of the current government. But, under central-planning socialism of the thoroughgoing variety being considered here, the government is, in effect, the only employer in the entire economy. This means that the government, ultimately, has final say in all matters of hiring, firing, and promotion, and on how resources are to be used throughout the economy. This will tend to preclude effective, public criticism of the government in two ways. First, even though, in this socialist system, there may be no legal barriers to criticizing the government, there will be practical barriers. If the government is one's employer, then any serious, public criticism of the government will leave one open to economic reprisals, to not getting promoted, to being transferred to an undesirable area, even to being fired and unable to get rehired anywhere else. As Leon Trotsky said, 'Where the sole employer is the state, opposition means death by slow starvation. The old principle, who does not work shall not eat, has been replaced by a new one: who does not obey shall not eat' (1937: 76). And merely to make economic reprisals illegal would not adequately solve the problem, since such reprisals, to be effective, need not be blatant enough to be discernible by legal authorities. In fact, to be effective they need hardly ever occur; the mere possibility of their occurring can be enough to discourage many people from any extensive, public criticism.

Moreover, the possibility of economic reprisals, and the intimidation it breeds, will extend to newspaper reporters, editorial writers, television commentators, and, in general, to those whose very job is to report upon, and critically evaluate, governmental activities. Thus, in central-planning socialism, a vigorous free press, reporting government activities impartially and criticizing the government without fear, will be unlikely. (And, of course, the evidence from actual socialist countries amply confirms this; see e.g. Evers 1989.)

In a capitalist economy, on the other hand, the government is far from being the only employer in the economy. There are many different employers; most, including those who employ newspaper reporters, editorial writers, and television commentators, are altogether independent of the government. Typically, capitalist employers are interested, first and foremost, in profits, not in promoting political ideologies; so, typically, they care only about their employees' productivity, not their political opinions. Thus, in capitalist democracies, people have little cause to fear economic reprisals for criticizing the government and will normally feel free to do so. (There are, of course, exceptions, such as the infamous McCarthy era in the United States; but these are indeed exceptions rather than the rule.)

A second way that the government's exclusive control over the economy constrains public criticism of the government is this. Often effective public criticism of the government requires access to certain resources: paper and printing presses for publishing books and pamphlets, radio and television time, auditoriums for speaking to large groups, and so on. But, in a centrally planned economy, all such resources are controlled by the government itself, and thus are available only with the government's consent. Obviously if one's reason for using these resources is to criticize the government, then this consent may well not be granted. To use one of Friedman's examples, say a person wants to publish the *Wall Street Journal* in a centrally planned economy. To do so, this person will have to convince the government of the value in doing so, and how valuable is the government likely to view a newspaper that is diametrically opposed to everything the government stands for? In short, the paper will remain unpublished. But now consider the parallel case of a person who wants to publish and distribute the socialist paper, *Daily Worker*, in a capitalist country. The government of this capitalist country will be no more impressed with the value of the *Daily Worker* than the socialist government was with the value of the *Wall Street Journal*. But since, in a capitalist country, the resources for publishing newspapers will not be controlled by the government, the government's views make no difference. Instead these resources will be controlled by private individuals who are interested primarily in making a profit. Therefore, provided only that the person can pay for them, these resources will be made available to her; she need not convince anyone of the newspaper's value—much less the very people who are vehemently opposed to everything for which the paper stands.

Of course, unless this person is independently wealthy, getting the funds to pay for these resources may well be difficult. But she may, as did Marx, find her Engels. Or she may form a corporation for publishing the newspaper and sell stock. The point is that, in a capitalist system, the potential sources of financial backing are many; if one source fails, a person can always try another; whereas in central-planning socialism, there is only one source—the government—and for projects critical of the government, this is an unlikely source indeed.

For these reasons we must, I think, agree with Milton Friedman: a thoroughgoing central-planning socialism is not as conducive to free speech, especially speech critical of the government, as is capitalism. Accordingly, such a system is not as conducive to political freedom. It is true that, in so far as one's political power in a democracy depends upon one's wealth (see Sect. 3.4 above), the grossly unequal distribution of wealth found in capitalism today mitigates its advantage with respect to political freedom. But to the extent that political power may be distributed more equally under socialism, this is largely a consequence of the mere fact that, as I have been trying to show, people are equally *without* adequate political power under socialism. We must conclude, therefore, that capitalism does in fact have the advantage here.

Finally, still another vital dimension of freedom is leisure, being free from the constraints of work. Some have argued that central-planning socialism might be more successful than capitalism in realizing leisure because of the competitive pressures of capitalism (see esp. G. A. Cohen 1978). The general idea is that these pressures drive people to work longer hours than is desirable, while central planners can build, into their plan, a better balance between work and leisure.

On the other hand, one might argue that the amount of work *vis-à-vis* leisure found in capitalism is, in part, a result of the choices made by people rather than government officials, and that people choosing for themselves are likely to arrive at a balance between work and leisure more favourable to themselves than would government officials. But then how many people really do choose for themselves even under capitalism? I suppose that those at least who are self-employed often are able to choose their own balance between work and leisure. Those who are not self-employed can perhaps express their choice politically, by voting for, say, those who support more state-sanctioned holidays or a governmentally imposed shorter working week which employers will be free to extend only by means of overtime wages. Moreover, if current capitalism turns out to be the more productive of the two systems, a matter we shall look into next, then, other things being equal, people may not need to work as many hours as they must in socialism in order to get by satisfactorily. As a matter of fact, the trend these days in many prosperous capitalist countries, such as Germany, is indeed towards shorter working weeks and longer vacations. In sum, I see no compelling reason for giving central-planning socialism an edge over current capitalism with respect to leisure; on the contrary, there may be reason for giving capitalism the edge, although this important matter deserves more consideration than I can give it here.

This completes my comparison of the two systems in terms of freedom. So which system then provides the most freedom overall? Freedom from the constraints of ignorance, disease, and certain injustices, and from social constraints, dimensions of freedom in which socialism has the edge, are probably more important than entrepreneurial, consumer, and maybe even political freedom, dimensions of freedom in which capitalism has the edge. This is a point in favour of socialism. On the other hand, the edge that socialism has over current capitalist systems in providing freedom from the constraints of ignorance, disease, and certain injustices is not, I think, very substantial, as current capitalist systems provide many of the government benefits necessary for these freedoms as well, and to the extent that they do not, they clearly could. But the edge that current capitalism has over central-planning socialism with respect to entre-preneurial, consumer, and political freedom is, I think, substantial, and necessarily so. That is to say, capitalism's edge in these areas is largely built into the nature of the systems themselves, and cannot therefore be significantly reduced without eliminating the very differences between the systems that, definitionally, differentiate them from each other—these differences being those between central planning and the market and between public and private ownership of capital goods. This is a point in favour of current capitalism. If, in addition to its significant edge in these dimensions, capitalism also has an edge, overall, in that dimension of freedom dependent upon wealth—that is, financial freedom—then I am prepared to say that current capitalism has the edge in freedom overall. Since, however, this dimension of freedom is largely dependent upon economic productivity, let us turn finally to a comparison of the two systems with respect to this value.

3.6 PRODUCTIVITY

To begin with, it is necessary to become clear about what exactly shall be meant by 'productivity' here. Is not productivity, it might be asked, simply gross national product (GNP) or, perhaps, net national product (NNP)? Not really. But then these concepts, especially that of net national product, are not altogether irrelevant either. Let me explain.

Consider, first, what GNP and NNP are. The gross national product of a nation is, roughly speaking, the market value of all the products—i.e. goods and services—produced in that nation in a given year. This includes, primarily, consumer goods, but also capital goods (investments) and governmental goods. But so as to avoid double counting, goods or services (such as wheat) that are not consumed directly, but instead contribute to the cost of goods that are consumed directly (such as bread), are not counted. The only exception to no double counting is new capital goods. New capital goods, although obviously not consumed directly, are nevertheless counted in GNP.

Net national product (NNP), on the other hand, is determined by deducting the total amount of depreciation upon all capital goods for the year from GNP. In this way, NNP more or less eliminates the double counting of capital goods that is built into GNP, thereby making it the more accurate measure of productivity. Nevertheless, because GNP is more easily determined than NNP, it is more widely used as a measure of productivity than NNP. But since NNP is the more accurate measure of productivity, it is the one that we shall focus on here.

'Real' NNP is the net national product corrected for inflation by stating it in terms of the prices of a base year. Finally, 'per capita' NNP is the net national product per person, which is the figure that can be used to compare the NNP of two nations with differing populations, or the same nation from one year to the next as its population varies. Whenever I mention NNP here, it is to be assumed that I am referring to real, per capita NNP.

So why not use NNP as the measure of a system's productivity? For many technical purposes it is, in fact, a useful measure of productivity. But for purposes of comparing the productivity of different economic systems, NNP simply will not do. Since NNP is calculated by using market prices, market prices would then become the standard by means of which productivity was measured. But market prices are, in many ways, distorted by market imperfections, so their accuracy as a measure of productivity leaves much to be desired. Among the most conspicuous of these imperfections are, as we have already seen, monopolies and externalities (both positive and negative). Monopolies (oligopolies, and the like) artificially inflate market prices, which makes it look as if there has been more productivity than there really was. Externalities are costs and benefits to third parties from production, or what is produced, that are not reflected in market prices. An important negative externality is pollution, the costs of which are a destroyed environment. Environmental costs, in effect, reduce productivity; yet since, typically, they are not reflected in market prices, this is another way in which market prices alone fail to provide an accurate measurement of productivity.

But what many, especially the opponents of capitalism, object to most about using market prices to measure productivity is this. Market prices, they say, do not accurately reflect people's *true* wants and needs. In capitalist systems, so it is argued, people's true wants and needs are corrupted by advertisements and other forms of capitalist brainwashing, thereby transforming them into false wants and needs; it is therefore these false wants and needs that market prices really reflect. And, as these critics correctly point out, if market prices do not accurately reflect people's true wants and needs, then market prices do not accurately reflect productivity.

This objection presupposes, of course, a clear distinction, at least in principle, between wants and needs that are 'true' and those that are 'false'. Although those who espouse this objection have never, as far as I can tell, formulated this distinction adequately, I suggest it can be adequately formulated as follows. Let us say that those wants and needs that count as people's 'true' ones are those the satisfaction of which are most in people's best interests or, in other words, most in their personal welfare. The tough part, of course, is to provide a defensible account of 'personal welfare'. But it is just such an account that I have tried to provide in Section 1.7. I shall assume, therefore, that the distinction between wants and needs that are 'true' and those that are 'false' is clear enough in principle.

Assuming then that this distinction makes sense, we must conclude that the critics of market prices are absolutely correct; actual market prices, and thus NNP, are far from ideal as a measure of productivity. An ideal measure would be based instead upon the prices that the economy's goods and services would have if everyone were fully aware of their true wants and needs, always purchased accordingly, and the market were free of imperfections such as monopolies, externalities, and so on.

What I claim, however, is this: in spite of the obvious shortcomings of actual market prices, and thus NNP, as a measure of productivity, NNP nevertheless can represent a useful approximation of productivity, provided appropriate adjustments are made. The key is the adjustments.

How exactly then should NNP be adjusted so that it represents a useful approximation of true productivity? This is a controversial matter that I cannot address here except briefly and in general terms. As already suggested, adjustments must be made for monopolies and externalities (both positive and negative).

Adjustments must also be made for certain inequalities of opportunity. Let me explain. If people receive a high income for doing what few others have the willingness or inherent capacities to do—brain surgery, for example—then the high income is justified; high prices that are a consequence of 'natural' scarcities such as these do, I submit, represent productivity. But a high income for some service is neither justified nor represents productivity to the extent that it is instead the consequence of an 'artificial' scarcity. An artificial scarcity in the number of people performing some service is one that results from *unnecessary* limitations upon people's opportunities to perform that service—limitations like those due to discrimination, or to needless restrictions upon the number of people allowed to get the necessary training. Let us refer to unnecessary limitations such as these as 'artificial' inequalities of opportunity. Artificial inequalities of opportunity are to be contrasted with 'natural' inequalities of opportunity, which are ones that either cannot be eliminated at all, or else cannot be eliminated except at too great a sacrifice of other values for their elimination to be justified. So, to the extent that the high income received for performing some service is a consequence of a natural inequality of opportunity, I claim the income is justified and represents productivity. To the extent, however, that the high income is merely a consequence of an artificial inequality of opportunity, it clearly is neither justified nor represents productivity. In principle, therefore, artificial inequalities of opportunity ought never to exist. In practice, however, many such inequalities have existed in the past, still do exist, and probably always will. But to the extent that they do exist, NNP must be adjusted accordingly.

And, significantly, adjustments must be made for goods and services that are self-produced and thus do not enter the market. I am referring to 'do-it-yourself' goods and services, such as food grown in personal gardens, home improvements on which only the owners themselves have laboured, and, most important of all, the vast amount of ordinary household cleaning and cooking chores that almost all people do for themselves. Self-produced goods and services are clearly part of a nation's productivity, yet since they never enter the market, they are not included in NNP. The remedy is somehow to impute market prices to all these self-produced goods and then increase NNP by the resulting amount.

Adjusting net national product so that it takes into account monopolies, externalities, artificial inequalities of opportunity, and self-produced goods is, of course, far from easy. How to do this accurately raises difficult technical, and philosophical, problems that should not be minimized. Consider, for example how to take into account the costs of what is no doubt the main negative externality of modern industrial economies, whether they be socialist or capitalist: namely, pollution and the destruction of the environment. How does one put a price upon the loss of open spaces, trees, clear lakes, and clean air? Or upon the disease, suffering, and loss of life resulting from this pollution? These are deep questions that I shall not pursue here, but that must not be overlooked. Economists have at least attempted to make adjustments in NNP for such things as externalities and self-produced goods, although with no more than limited success. One of the best-known attempts is that by William Nordhaus and James Tobin (1972). The resulting figure is referred to by them as not net national product, but 'net economic welfare' (NEW).

In addition to adjustments to NNP for externalities and self-produced goods, Nordhaus and Tobin also include a monetary adjustment for the amount of leisure time that people throughout the economy enjoy. Although leisure time is clearly a crucial component in overall welfare, I prefer not to include it as a component of productivity. We must be on our guard against trying to pack too much into the concept of 'productivity'. If we pack too much into this concept, it will simply collapse into the concept of the 'general welfare'. And by collapsing into the concept of the general welfare, 'productivity' will then lose its usefulness as one of the more specific, prima-facie values that go to make up the general welfare, and that therefore can serve as a useful guide for determining what is in the general welfare (as explained in Sect. 2.2 above.) Rather than include leisure as a component of 'productivity', it is, I suggest, more appropriate—that is, more compatible with our intuitive understanding of the values in question—to include it as a component instead of 'freedom' in the broad sense (see Sect. 3.5).

So, in conclusion, 'productivity', as I am using the term here, is roughly equivalent to real, per capita NNP, adjusted (at least) so as to take into account monopolies, externalities, artificial inequalities of opportunity, and self-produced goods. If every appropriate adjustment to NNP had been taken into account and calculated perfectly, then the resulting figure would represent NNP as determined by 'ideal' market prices, ones that reflected true wants and needs. Such a figure would delineate an economy's productivity as precisely as possible. But, of course, such precision is impossible. Yet if we take into account as many appropriate adjustments as we can and calculate them as accurately as we are able to, then the resulting figure should at least represent a reasonable approximation of productivity. Since productivity, so understood, is to be measured in terms of (adjusted) market prices, this raises the obvious question of how the productivity of a centrally planned economy can, for all practical purposes, be determined, since such an economy does not give rise to market prices. The answer is that the goods and services generated by the centrally planned economy in question must have a reasonably accurate market value imputed to them. Economists have ways of attempting to do this—although exactly how to do this is a somewhat controversial and technical matter into which we need not enter here.

We now have before us a rough explication of 'productivity'. A meaningful comparison of the productivity of socialism and capitalism must, of course, be one that presupposes that both systems are starting with the exact same set of resources. And if we take 'efficiency' to be equivalent to productivity relative to a given set of resources, then it follows that the most productive system will be the same as the most efficient one. Let us now proceed with the comparison.

We may begin by acknowledging a number of significant advantages with respect to productivity that central-planning socialism has. First, as we have already seen, with central planning it is possible to maintain full employment all of the time. People's jobs are not, as in capitalism, dependent on fluctuations in the market. And socialism is not subject to the periodic recessions that plague capitalism, temporarily reducing productivity and aggravating the problem of unemployment.

In considering the contribution to productivity in full employment under socialism, however, we must keep this in mind. Full employment under socialism will inevitably be achieved only by employing a number of people whose marginal product—i.e. whose contribution to productivity—falls short of their wages. Since such people are paid an amount greater than their contribution, they, in effect, decrease overall productivity. In reply to this it might be pointed out that, morally speaking, these people cannot simply be allowed to starve; society would have to support them anyway. And it is more productive for society to support them by giving them a job at which they at least contribute something, rather than simply placing them on welfare without a job—as is typically done in capitalist systems. Perhaps so; but if this is about all that the gain in productivity from full employment under socialism amounts to, then any such gain may be modest.

With control over prices, central planners can, of course, suppress inflation. Actually, however, being able to suppress inflation merely through heavy-handed price controls is no real advantage. Generally speaking, such controls will serve only to cause inflation to surface again in other, equally unpleasant, guises, such as serious shortages, long waits in lines for scarce goods, and black markets.

A more genuine advantage of socialism is that, as we have seen, central planners have no reason for 'artificially' limiting the productivity of certain goods and services so as to raise their prices, as do those with more or less monopolistic control over certain goods and services in capitalism. Obviously the less productivity is thus limited, the greater productivity will be. It is very debatable how much productivity under capitalism is in fact lost due to monopolistic behaviour; perhaps the overall amount is not extremely great (Scherer 1980). But that some productivity is thus lost is undeniable.

Other sources of reduced productivity found in capitalism, but not in central-planning socialism, include various kinds of business strategies falling under the general heading of 'planned obsolescence'. As we saw, these strategies include not only purposefully designing products so that they wear out sooner than necessary; they also include the introduction of improvements in existing products, such as automobiles, TVs, and computers, at a slower rate than necessary, and deliberately manipulating, at regular intervals, the public's conception of fashion.

The best remedy against planned obsolescence is a business community with integrity. The next best remedy—but one that may be a more likely possibility—is healthy competition and informed consumers. The more healthy the competition and informed the consumer, the less likely it is that planned obsolescence strategies will work. I suspect that competition in current capitalism is generally healthy enough, and consumers well enough informed, so that, although planned obsolescence does detract from productivity, it does not do so enormously.

Unemployment, recessions, monopolies, and planned obsolescence are all well-recognized sources of underproductivity for capitalism. But what is probably the most serious source of underproductivity for capitalism is one that is less widely recognized: namely, lack of opportunities. As explained earlier (Sect. 2.6), productivity requires not only that people have the appropriate incentives to be productive, but also that they have the appropriate opportunities. Poverty, inadequate education and medical care, and vast inequality in the distribution of wealth all contribute, as we have seen, to a lack of opportunity for many under capitalism. Greater equality of opportunity will result in more overall realization of human potential, and greater compatibility between people's inherent capacities and their occupations—all of which will mean greater productivity. Since, for reasons we have already seen, the version of socialism being considered here succeeds in realizing equal opportunity more fully than does current capitalism, to this extent it will be more productive.

Other advantages in the area of productivity have been claimed for socialism; these include happier, less alienated workers, better control over negative externalities, and less underproductivity resulting from positive externalities. I think, however, that I have touched upon socialism's most defensible claims already. Workers under socialism, with its authoritarian central planning, will certainly have no more freedom and autonomy than workers under capitalism; indeed, with everyone working for the state, they will not even have a choice among ultimate employers (cf. Gray 1989: sect. iv). So it is unlikely that workers under socialism will be much happier, or less alienated. And any serious problems that externalities cause can, in principle, be, and often are remedied under capitalism by appropriate state intervention. Appropriate state intervention may at times be difficult to come by under capitalism, owing to shameless resistance from private enterprises motivated only by profits and unscrupulous legislators who cater to special interests. On the other hand, the pressure to meet quotas under socialism has contributed to much shameless pollution as well and, as we shall soon see, socialist legislators will hardly be free from special interest influences themselves, not to mention pressure from voters to take a short-term perspective. Therefore any difference here between the two systems may well be of little significance.

Let us turn now to the main productivity advantages of capitalism. The first advantage of capitalism is simply its consumer freedom. At least if we assume a distribution of dollar votes that is not excessively unequal (with what is meant by 'excessive' here having been explained in Sect. 3.5), capitalism, with its consumer freedom, will be more successful than central-planning socialism in generating goods and services that reflect people's true wants and needs. As explained above, what I mean by people's 'true' wants and needs are those the satisfaction of which are most in people's best interests or personal welfare. And in Section 1.7, I attempt to provide a satisfactory explication of 'personal welfare'. For convenience, let us refer to those goods and services that satisfy true wants and needs simply as the 'right' or 'correct' ones. What I am claiming is that, at least if we assume a distribution of dollar votes that is not excessively unequal, capitalism will be more successful than central-planning socialism in generating the right goods and services. This is largely because, as we have seen, the knowledge and information that central planners have can never match the combined knowledge and information of the many thousands of people who, through free-market competition, generate the goods and services of capitalism.

Let me explain this further with the help of an analogy. Consider, for a moment, how best to arrive at what is correct, not with respect to goods and services, but beliefs. What, in other words, is the best way for the truth to emerge? Is it through governmental censorship, or through free competition among all beliefs for people's allegiance? Surely, as John Stuart Mill pointed out (Mill 1859), it is through free competition among all beliefs; in other words, through free speech rather than censorship. For one thing, the knowledge and information that government censors have at their disposal for trying to determine what is true can never match the combined knowledge and information of the people as a whole. The knowledge and information of the people as a whole is, of course, widely dispersed among many individuals. Some of these individuals, being misguided and uninformed, will therefore believe what is false. But others, with more extensive knowledge and information, will believe what is true. And it is their beliefs, the true ones, that are likely to prevail in the long run since, after all, these beliefs enter the competition for people's allegiance with a decisive advantage: their truth.

Likewise, I claim, the best way for 'correct' goods and services to emerge—that is, those that reflect people's true wants and desires—is not through the government selecting what goods and services to produce and in what quantities, but through free competition for people's allegiance among all goods and services; that is, through the free market, not central planning. The knowledge and information that central planners have to aid them in selecting goods and services can never match the combined knowlege and information of the people as a whole. Once again, the knowledge and information of the people as a whole is widely dispersed among many individuals. Some of these individuals, being misguided, perhaps by deceptive advertising, will inevitably choose goods and services that do not reflect their true wants and needs. But others, with more knowledge and information, will choose the right ones, those that do reflect their true wants and needs. And, with a free market, it is the right goods and services that are likely to prevail in the long run since, after all, they have a decisive advantage: they are the ones that (by hypothesis) really do reflect people's true wants and needs.

I do not mean to suggest by this that free market competition be made a matter of right, as is free speech. This is because—in contrast to *some* censorship of ideas and opinions—*some* governmental regulation of the free market is justified. I have in mind here governmental regulation of monopolies, externalities, and dangerous products. But, in spite of these exceptions, free-market competition remains the best way, generally, for the right goods and services to emerge. Just like government censors cannot match the success of free speech in generating true beliefs, central planners cannot match the success of the free market in generating goods and services that reflect true wants and needs. And, of course, the more successful an economy is in generating goods and services that reflect true wants and needs, the more productive it will be. An economy cannot be productive merely by turning out buggy whips, so to speak, no matter how many millions it may turn out.

Take, for example, medical services. Say that the central planners got lucky and were able to hit upon just the right quota—that is, *quantity*—of physicians called for by people's true wants and needs. The central planners would still need to determine the right degree of *quality*. Let me explain.

According to the version of central-planning socialism being considered here, for each occupation throughout society the central planners must determine the quantity of people that are to pursue that occupation; let us refer to this quantity as the occupation's 'quota'. And, as explained, once the central planners set the quota for an occupation, they must then try to make sure that salaries have been set so that the quota is filled. If more people are attracted to some occupation than there are positions available, then the quota is to be filled by choosing the most qualified among them.

Consider now the occupation of 'professional athlete', an occupation that I assume will be available under socialism just as it is under capitalism. Other things being equal, this occupation will be very popular; it is far more fun to play sports than to dig ditches or work in a factory. But if the salary for those pursuing this occupation were so low that it provided for no more than bare subsistence, while the salaries for virtually all other occupations were much higher, some very much higher, then clearly, in spite of the inherent attractiveness of this occupation, few qualified people would choose to pursue it, so few, probably, that the quota would barely be met. Most potentially good athletes would choose to pursue some other, higher paying occupation instead. What this would mean is that the level of performance in professional sports would be minimal. If, however, the salary were to be set at an amount far higher than that for most occupations, then the competition for the limited number of positions available would be fierce, with only the very best athletes having any chance of landing a position. The level of performance in professional sports would then be superb. Those who lost out in this competition to become athletes would have to fall back upon some other occupation, perhaps one the salary for which would be lower than that necessary for filling the occupation's quota except for the fact that many people unable to get anything else end up there.

The point is simply this: the higher the central planners set the salary for some occupation above the bare minimum necessary for filling its quota, the higher the average level of performance in that occupation will be. So, in setting the salaries for each occupation, the central planners will, in effect, be setting the relative levels of performance in these different occupations as well. I assume that they should try to get exceptionally high levels of performance in certain occupations such as brain surgeon, research scientist, and so on; mediocre levels of performance in these occupations would be serious indeed. But to get exceptionally high levels of performance in these occupations, the planners must do either one of two things: either they must set the *salaries* for those in these occupations high enough to stimulate such extensive competition for the available positions that only exceptionally qualified people will be chosen, or they must set the *minimal qualifications* for those in these occupations so high that only exceptionally qualified people can meet these qualifications. By going the route of unusually stringent minimal qualifications, the central planners will then have the additional burden of formulating these qualifications in a clear and meaningful way. Moreover, by going this route they will probably not avoid the necessity of setting unusually high salaries for those in these occupations anyway, since such salaries will normally be necessary just to flush out the relatively few who can meet the unusually stringent minimal qualifications.

All of this, however, will mean more problems for the central planners. Unusually high salaries for some occupations will, of course, tend to undermine one of the main attractions of central-planning socialism: its relatively equal distribution of income. Even more significant, however, is this. Not only will central planners have to choose the appropriate quantity of people in every occupation, but they will also, in effect, have to choose the appropriate level of performance for every occupation relative to that for every other occupation. Yet to do all of this correctly requires information that, by and large, central planners cannot possibly possess, information about what quantities and relative levels of performance would most accurately reflect people's true wants and needs. Thus many of their decisions will, inevitably, be incorrect, and to the extent that their decisions are in fact incorrect then, of course, productivity suffers.

In capitalism, on the other hand, the quantities of people in each occupation, the levels of performance, and the differentials in salaries are all 'decided' simply by means of a free, competitive market which, as we have seen, is more likely to reflect people's true wants and needs than is central planning. And to the extent that the results of a free, competitive market do more accurately reflect people's true wants and needs then, once again, to that extent capitalism will be more productive.

So far I have been concentrating on the lack of congruity in central-planning socialism between the goods and services that will be produced and those that will satisfy people's true wants and needs. There is, however, another congruity problem with central-planning socialism: that of *internal* congruity, of having all the innumerable goods and services turned out by the economy fit together well with one another so that serious shortages and surpluses do not arise. Given the hundreds of thousands of different kinds of goods and services necessary for a modern economy—automobile repair services, microchips, grade X rubber, 6-volt batteries, computer programming, cloxacillin, and so on, and so on—co-ordinating everything adequately so that it all fits together without serious shortages of surpluses is no small problem. The mathematical techniques for doing so are available, as is the necessary computer power, but will the central planners be able to accumulate the necessary empirical information to plug into their equations and feed into their computers?

In a free-market, capitalist economy, of course, everything fits together automatically by means of supply and demand. No one producer or distributor in a free market has enough information to work out the proper fit among goods and services for the entire economy, but each will, typically, have enough information to make things fit together well enough in his or her own limited domain of production or distribution. As Allen

Buchanan puts it, the market 'can be viewed as a device for effectively coordinating the actions of many individuals through specialization in the gathering and use of information' (1985: 17). In order for things to fit together in a centrally planned economy as adequately as in a free-market economy, the central planners would need an amount of information approximating that which exists in a free-market economy, not in any one place, but dispersed among these many thousands of independent producers and distributors. That this enormous amount of information could ever be collected in one place and made available to the central planners is highly unlikely. Keep in mind that there are mind-boggling problems not only with respect to collecting and properly aggregating into manageable form all this information, but even with respect to motivating people to report it accurately in the first place. If, for example, a manager's rewards depend upon whether or not his unit is able to meet its quota, then he has an incentive to understate his unit's productive potential in the hopes of getting back from the planners a relatively low, easy to fill quota.

And, in any case, circumstances are always changing—the wants and needs of people shift continuously, technological breakthroughs occur, natural disasters happen (droughts, floods) creating unexpected shortages, and so on. With each such change of any significance, the process of information-gathering will have to begin again, and an entirely new plan, fitting everything together differently so as to accommodate this change, will have to be drawn up. So even if all the necessary information could somehow be accumulated in one place (which is doubtful), this information would, because of constantly changing circumstances, almost immediately become obsolete anyway, probably even before a long-range plan based upon it could be drawn up (for more detail, see Mises 1951; Hayek 1935, 1967; Arnold 1989, 1990). Accordingly, it is doubtful indeed that this version of central-planning socialism can ever achieve the degree of internal congruity usually found in capitalism. And, of course, to the extent that it does not, this system will, once again, be less productive.

To be sure, capitalism suffers from its own sort of incongruity, an incongruity that takes the form of periodic slowdowns in the economy giving rise to serious unemployment. And the information that business leaders have under capitalism is less than perfect too (see e.g. O'Neill 1989). But the information problems facing them are minor compared with those facing central planners. Moreover, means for combating the internal incongruity of capitalism are available. These take the form not just of government monetary and fiscal policies, but also various programmes for combating unemployment directly that have yet to be fully utilized (see Sect. 5.7 for some details). And the informational problems of capitalism can be mitigated by more government collection, collation, interpretation, and distribution to business leaders of certain information about the economy as a whole, about the extent of new investment plans in different areas, and so on.

Means for overcoming the much greater information problems of central-planning socialism are less promising. Ingenious proposals have been made for overcoming these problems by combining aspects of a market economy with state ownership and control over all investments and capital goods (e.g. Lange 1964; Lerner 1946). I shall, in the next chapter, consider one such proposal in detail, that put forth by David Schweickart and called by him 'worker control socialism' (Schweickart 1980). But, for now, I have just two observations to make. First, to the extent that these proposals succeed in overcoming the information problems of central-planning socialism, they do so only by abandoning this system in favour of a system, sometimes called 'market socialism', that is closer to capitalism. Secondly, regardless of how well these proposals overcome the information problems (Shapiro (1989) argues that the Lange–Lerner proposal fails to do so), there remains still another problem regarding productivity under central-planning socialism that these proposals do not overcome: the incentive problem. Let us turn now to this problem.

Free markets serve two productivity functions. Not only do they provide producers and distributors with the information necessary for being productive; they also provide them with the incentive. Their very livelihood depends upon their being more productive than their competitors. To the extent that they are productive, they personally reap the profits. To the extent that they are not, they personally suffer the losses. In central-planning socialism, however, the state owns all capital goods, makes all investment decisions, reaps any profits, and suffers any losses. Producers and distributors are, for the most part, merely to be paid a salary by the state sufficient to entice them to work. Any major innovations or new directions for their enterprises that these producers and distributors come up with cannot be put into effect without government approval, which, of course, may not be forthcoming. And even if they do get the approval, they will not get the profits. Without their personally getting the profits from their efforts or innovations, and without any assurance that their innovations will even be put into effect, their incentive to work hard and be innovative cannot be expected to be as great as it would be with capitalism. And, of course, to the extent that their incentive to work hard and be innovative is not as great, productivity will suffer.

Special bonuses under socialism for being unusually productive or innovative will mitigate the incentive problem somewhat, but only somewhat. Bonuses are usually rather limited in amount and cannot possibly permeate the economy as extensively as the profit motive does under capitalism. Moreover, the potential of bonuses for motivating technological innovation may be undercut somewhat by the fact that the system also encourages a steady flow of output, which tends to discourage any disruptions that may be necessary for experimentation and technological innovative. (And there is also the so-called problem of 'success indicators'—that is, the problem of how to define exactly what target criteria are to be met in order for one to be eligible for a bonus. See Nove 1962.)

Next, consider something that advocates of central-planning socialism claim is one of its advantages, but that, I shall argue, is actually an advantage of capitalism's: namely, a capacity for economic growth. Growth is not the same thing as productivity, just as income is not the same thing as wealth. But obviously the relationship between growth and productivity is very close; growth is the basis for future productivity. Therefore it is not inappropriate to consider a system's growth capacity in evaluating how successful it will be, overall, in realizing the ideal of productivity.

Advocates of central-planning socialism argue that central planners, having full control over the economy, can always orientate it towards the production of capital goods—that is, new investments—to a greater degree than, in a free market, will occur naturally with supply and demand. And it is largely upon the production of new capital goods—steel mills, power plants, and so on—that economic growth depends. Thus, it will be concluded, central-planning socialism has a greater capacity for economic growth which, of course, will be a crucial advantage, in the long run, in realizing the ideal of productivity.

In reply to this argument, the first thing to be said is that capitalist governments have the tools for bringing about a higher rate of investment than would occur naturally as well, if a higher rate is deemed desirable. These tools are, to be sure, more indirect than those available to central planners, but they can be effective nevertheless. Included among these tools are monetary and fiscal policies, and other, somewhat less conventional policies such as the ones mentioned briefly in Section 6.4.

The next thing to be said is that, not only does the socialist's argument understate the capacity of capitalist governments to bring about higher investment rates than would occur naturally, but it also overstates the capacity of central planners to do so. Capital goods, upon which economic growth depends, include not just material capital, but human and technological capital as well. Human capital is in the form of the talents and abilities that people in the society possess, and technological capital is technological know-how. Both human and technological capital depend in large part upon people's incentives to develop their own talents and abilities, and to develop new ideas and improved technologies. But, with people not being able to control and profit personally from their own talents and abilities or new ideas and improved technologies, central-planning socialism, as compared with capitalism, might well be deficient in motivating people to develop them. And to that extent economic growth would suffer.

Moreover, as with a centrally planned economy in general, centrally planned investment presents the danger of excessive paternalism—paternalism that goes well beyond any reasonable guideline. If the distribution of wealth is not exceedingly unequal, then the people themselves, through their 'dollar votes' are likely to be more successful in choosing goods and services that reflect their 'true' wants and needs than are central planners paternalistically choosing for them. Even if (as would be unlikely) the central planners did reject paternalistic investment and tried instead to respond as accurately as possible to people's wants and needs as seen by the people themselves, these planners could never get the knowledge and information necessary to respond as accurately as does free market investment. And to the extent that, as a result, central planners end up investing in buggy whips, so to speak, growth will suffer.

Another potential problem is a consequence of the fact that the version of central-planning socialism being considered here is coupled with a Western-style democratic political system. Every time a new political party that is going in a different direction from the old one takes control of the central planning, the momentum of growth that the old party had established in certain areas may be cut short prematurely for the sake of establishing momentum in different areas, thereby slowing down the overall rate of growth, at least temporarily. Since, in capitalism, the economy is much more independent of politics, changing the party in control is unlikely to have a similar disruptive effect upon economic growth.

But these reasons why the growth potential of central-planning socialism is not as great as socialists often claim are well-known. I want now to present some reasons that may be less well-known, reasons involving, once again, the fact that the socialist system being considered here is coupled with a democratic political system. Rapid economic growth can, of course, be achieved only by directing significant amounts of a nation's resources away from consumer goods into capital goods; it can be achieved, in other words, only at the expense of considerable immediate sacrifices from the general public. But, as those in Western democratic nations know only too well, in proposing immediate sacrifices from the public for the sake of long-term benefits, politicians can go only so far and still hope to get elected. This has been shown by recent American presidential elections in which the slightest hint by a candidate that he might tolerate an increase in taxes has doomed him from the start even though, given the enormous national deficit, virtually all reputable economists agreed that some increase was most in the nation's long-term interests. The voting public simply does not normally have a very long-term perspective. The public can at times be persuaded to accept moderate sacrifices for the sake of the long run, but—short of a perfectly obvious and dire national emergency, such as being attacked by a foreign power—the general public normally cannot be persuaded to accept sacrifices of the scope in question here. To extract such sacrifices from the public in relatively normal times, it generally takes a totalitarian government, one not very responsive to the will of the people—a government like the former Soviet Union and other central-planning socialist countries had until recently. In sum, central planners answerable to the general public through democratic elections must act only within the parameters of what is politically feasible, and substantial cuts in consumer goods such as we are talking about here may not normally be politically feasible. Indeed, it might, at times, be politically difficult for these central planners even to avoid *negative* rates of investment, rates at which current capital was being consumed.

But the greatest problem with democracy is not the shortsightedness of the electorate, but the extent to which it breeds governmental support for special interests. 'Special' interests are group interests that are neither in the general welfare, nor protected by any rights. Government support of the special interests of some group therefore amounts to no more than governmental bias in favour of this group. Take, for example, support for the merchant marine industry by subsidies for shipbuilding and other operations, and by certain restrictions on foreign coastal traffic. As Milton Friedman points out (1980: 292–3), it is estimated that (as of around 1979) US government programmes in support of this special interest cost American taxpayers $600 million a year, which amounted to a $15,000 per year subsidy for each of the 40,000 people actively engaged in this industry. Other examples of government support of special interests in blatant disregard of the general welfare include most protective tariffs and quotas on imports (see e.g. Blinder 1987: ch. 4; Friedman and Friedman 1980: ch. 2), and the scandalous deregulation of the US savings and load industry that, it is currently estimated, may cost US taxpayers from $330 to $500 billion over the next decade alone (Waldman 1990).

Conservatives, such as Friedman, are very sensitive indeed to special interests that are supported by governmental activity. But they are absurdly insensitive to special interests that are supported by governmental *in*activity. Yet government officials, through failing to act, can be just as crass in disregarding the general welfare as they can through acting. Take, for example, the government's failure for seven years, at the behest of the Ford Motor Company, to pass regulations that would have prevented a number of people from burning to death each year as a result of the Ford Pinto's faulty petrol-tank design (Dowie 1977).

Why do special interests often succeed in gaining governmental support even when they are so clearly contrary to the general welfare? The reasons are varied and complex, and include the skill of professional lobbyists in the use of unscrupulous stalling techniques, and the shameful pro-industry bias of many government bureaucrats who, having been recruited from the very industries they are supposed to regulate, feel a greater loyalty to their former business colleagues than to the public they are supposed to be serving. But probably the greatest reason is that, all too often, a politician's support of special interests translates into more votes, or campaign funds, than would support of the general welfare. It is not difficult to see why this should be so. Take, again, the blatant governmental favouritism towards the merchant marine industry. Since the governmental programmes favouring this industry were (around 1979) worth an average of $15,000 per year for each of the 40,000 people in this industry, not only were each of these 40,000 people thus likely to vote for the politicians who favoured these programmes, but they also would have found it worthwhile to contribute significant funds for the election of these politicians. The gain from, instead, discontinuing these programmes would have been $600 million a year, which would have benefited everyone in the United States other than those in this industry, but in an amount equal to only about $3 per person, per year. Few people are going to vote, or not vote, for a politician merely for the sake of $3. Thus the public, in effect, gave these politicians a choice between 40,000 extra votes and fat campaign funds for supporting these programmes, and virtually nothing for supporting instead the general welfare. Which were politicians likely to choose? There are no doubt a few genuine statesmen who would have sacrificed the votes and funds for the sake of the general welfare. But genuine statesmen are all too rare and, obviously, tend not to get elected. In short, the general welfare is often diffused over so many people that no single person's stake is great enough to influence his or her vote or inspire campaign contributions, while special interests are concentrated upon so few people that, consequently, each of their stakes in the outcome is enormous, thereby guaranteeing their votes and campaign contributions.

Normally, no one special interest will have the power to deliver enough votes or campaign funds by itself to get a politician elected. But many an unscrupulous politician gets enough votes or funds to be elected by catering to a number of such special interests at once. As Alan Blinder puts it: 'In politics, the principle of reelection overwhelms the principles of both equity and efficiency' (1987: 122). For these reasons, so many special interests prevail that even those who thereby benefit the most end up losing more than they gain—these losses being in the form of higher taxes and dangerous products resulting from governmental support of *other* people's special interests. Thus everyone would be better off by giving up the benefits from governmental support of their own special interests in return for everyone else doing likewise. But, of course, the 'assurance' problem (Sect. 2.5) prevents this from ever happening. So those who live in a democracy continue to be plagued by governmental support of special interests in crass disregard of the general welfare.

How then does all of this relate to the comparison between socialism and capitalism with respect to growth and productivity? First of all, since the version of socialism being considered here is coupled with a democratic political system it will, of course, be plagued by governmental support of special interests as well. To be sure, the motive that industries and professional groups have for seeking governmental favours under socialism will not be higher profits, as there are no private profits under socialism. Rather their motives will include such things as higher salaries, more leisure, and better working conditions relative to others in the economy. But then with politicians all competing with one another in promising favours for special interest groups—as the election pressures of a democracy inevitably drive them to do—it will become more and more difficult for central planners to maintain an adequate rate of investment and growth, for as each of these promises is fulfilled, the amount of money left over for investment and growth will become less and less. And not only will this version of central-planning socialism be plagued with special interest problems, but these problems are likely to be even greater than with capitalism. Certain countervailing features of capitalist democracies— features that help keep government support of special interests within tolerable bounds—will be less prominent with central-planning socialism.

The first of these features is that, in capitalism, the economy is not, by and large, run by the government. Therefore, in capitalism, the opportunities for government support of special interests throughout the economy are naturally less extensive than in central-planning socialism where the government does run the economy. Socialists, with some justification, are accustomed to denouncing capitalism for the monopolies it generates. But central-planning socialism can be viewed as being one enormous monopoly itself, one in which the monopolist is the government, and the monopoly extends not over just one industry, but over the entire economy. It is not, to be sure, a monopoly dedicated to maximizing profits for capitalists; rather, it is a monopoly dedicated largely to maximizing power and influence for governmental officials, thereby enabling them to remain in control. And, of course, the way government officials in a democracy often go about maximizing their power and influence is through support of special interests. But a monopoly that achieves its goals through support of special interests can undermine productivity and growth just as surely as can a monopoly that—capitalist style—achieves its goals through artificially restricting production and raising prices.

Secondly, the fact that, in capitalist democracies, much special-interest legislation must come up for a vote piecemeal—or combined with just a relatively few other measures—helps constrain special interests somewhat, since then legislators supporting the special interests must somehow get the support of legislators with constituencies that do not benefit from the legislation. In central-planning socialism, however, the central economic plan will have to be voted on more or less as a whole, since if legislators were able to tinker around with it too much, the plan would soon become hopelessly incongruent. But then, if the plan must be voted upon more or less as a whole, there will be little opportunity for legislators to defeat those special-interest provisions buried in the plan that do not benefit their own constituents. In other words, the central economic plan becomes an effective vehicle, one that does not exist in capitalism, for slipping special-interest provisions by legislators virtually unchallenged.

Finally, in capitalist democracies a free and vigorous press along with the general public are often quick to openly criticize governmental support of special interests. These criticisms help counter any gains that politicians may derive from supporting these interests, and thus act as a healthy constraint. But, as we have seen (Sect. 3.5), it is just such healthy criticism of the government that central-planning socialism is likely to inhibit, since it is upon the government that all jobs, promotions, raises, and so on, ultimately depend. So to the disadvantages of socialism with respect to productivity and growth we must add that of excessive catering to special interests.

Overall, a central-planning socialist system tends not, for these various reasons, to work well when combined with a Western-style, democratic political system. Indeed, democracy and central-planning socialism may well be inherently incompatible.

So, it may be asked, why not give up combining the two; why not simply combine central-planning socialism with a totalitarian political system instead? This, unfortunately, is no solution. Democracy does, to be sure, have its serious problems, the most serious of which is the frenzied pursuit, within the political arena, of selfish, special interests. It would be a wonderful world if, within the political arena, most politicians, business executives, and citizens instead felt morally obligated to pursue what they sincerely believed to be in the general welfare, and did so (cf. David Miller's (1989) distinction between politics as mere 'interest-aggregation' and politics as 'dialogue', ch. 10). In the real world, however, bigotry, greed, and ignorance, along with the enormous assurance problems inherent in democratic procedures such as secret balloting, are major obstacles to this happening. Yet as serious as the problems of democracy are, the problems of totalitarianism are even worse. Take, in particular, special interests. Not even dictators can retain power without keeping certain groups happy—especially those groups controlling the use of force.

Therefore dictators must cater to special interests no less than do politicians in a democracy. But with a totalitarian system the problem of special interests is actually enhanced, since the very things that, with a democracy, normally keep the pursuit of special interests more or less within tolerable bounds—namely, free speech, free press, open debate by public officials, and democratic checks and balances—will, with a totalitarian system, be largely absent. In short, the pursuit of special interests in a totalitarian system is likely to make the special-interest problem of democracy look trivial by comparison.

Given the absolute control that the government in a country with central-planning socialism has over the entire economy, it is especially crucial that the political system of such a country succeed in keeping the government from straying too far from the general welfare towards special interests. The most successful political system humanity has yet devised for keeping the government from straying too far from the general welfare is a healthy democracy. So if, as I have suggested, central-planning socialism is incompatible with a healthy democracy, then this is a serious flaw indeed.

This then completes my comparison of the two economic systems with respect to productivity. Although central-planning socialism appears to have some genuine advantages with respect to productivity, these advantages depend largely upon the presence of a healthy democracy to keep the government from straying too far from the general welfare. But, as I have argued, a healthy democracy is exactly what such a system is likely to lack. And, in any case, productivity will be constrained by serious information and incentive problems that are very largely avoided by a market economy such as that of current capitalism. I conclude therefore that, overall, current capitalism has a significant edge with respect to productivity.

3.7 TWO CONCLUSIONS

From the above comparison of central-planning socialism with current capitalism, I am led to conclude two things. First, as far as I am concerned, neither system is worthy of our support. This comparison has revealed that both systems suffer from grave moral deficiencies. Current capitalism is deficient primarily with respect to the ideals of equal opportunity and equal access to the necessities for a decent life. Central-planning socialism is deficient primarily with respect to the ideals of freedom and productivity. Each system, of course, has its strengths as well; the strengths of each correspond largely to the deficiencies of the other. So, for a system worthy of our support, we must look for some new, or extensively revised, system that succeeds in overcoming the deficiencies of one of these systems while preserving its strengths.

This brings me to the second conclusion. Any new, or revised, system that succeeds in doing this will have to be one that, for the most part, features a free market rather than central planning. Although both central-planning socialism and current capitalism have grave deficiencies, comparing the two reveals that the deficiencies of central-planning socialism are more recalcitrant, more an inherent part of the system itself. In short, these deficiencies cannot be remedied without retreating considerably from thoroughgoing central planning, a retreat that would amount to almost a complete surrender to the market. The deficiencies of current capitalism, on the other hand, can, I think, be overcome without retreating, considerably, from the market and its characteristic strengths.

But how? In the remaining three chapters I shall try to show how. The proposals made in these three chapters, taken together, delineate an economic system that just may be worthy of our support.

4
WORKER CONTROL

4.1 INTRODUCTION

With capitalism, ordinary working people—assembly-line workers, low-level, dead-end white-collar workers, manual labourers, hired sales clerks, and so on—must bear a triple burden. First, as compared with those in more glamorous occupations—doctors, lawyers, engineers, executives, entrepreneurs, professional athletes, entertainers—ordinary working people are paid very little. Secondly, ordinary working people have relatively little social prestige. Third, as compared with those in more glamorous occupations, ordinary working people suffer from work that offers little satisfaction. The work that those in more glamorous occupations do, although difficult, is nevertheless often exciting, challenging, even fun and, in general, personally satisfying. But the work that ordinary working people do is exhausting, often dirty, almost always excruciatingly boring, and, in general, unsatisfying. In fifteen years of research on job satisfaction, Stanislav Kasl (1977) found that the conditions conducive to the low job satisfaction of ordinary working people include a lack of control over work; inability to use skills and capacities; highly fractionated, repetitive tasks involving few diverse operations; and no participation in decision-making (Kasl 1977; for more on job satisfaction, including a discussion of the accuracy of Gallup polls on 'satisfaction at work', see Kahn 1972). Many years of working under these conditions may even have an effect upon mental health. A study of automobile assembly-line workers, for example, found that about 40 per cent suffered from a significant mental-health problem, and only 18 per cent could be said to have 'good mental health' (Kornhauser 1965).

Since many people seem not to recognize something that should go without saying, I will say it. Without perfectly equal opportunities for everyone, which we know there will never be—without, that is, perfectly equal inborn capacities, perfectly equal access to education, health care, and other necessities, perfectly equal wealth, perfectly equal guidance from parents, perfectly equal freedom from discrimination, and even perfectly equal luck—the prerequisite for everyone's having a genuinely free choice of work will be lacking. And without a genuinely free choice, those who endure the least satisfying work do so largely because they must do so, not because they have chosen to do so.

In a perfectly just universe, it would seem that those who had to endure the least satisfying work would, as partial compensation, at least get the most pay or the most prestige. Yet, under capitalism today, those who have to endure the least satisfying work also get the least pay and the least prestige. In short, they get the worst of everything. No economic system can, of course, eliminate dull, unsatisfying work. But perhaps some system can at least make the jobs of those who must do this work more worthwhile than at present.

Worker control is designed to be just such a system, a system that mitigates the triple burden of ordinary working people as much as possible, and does so without sacrificing the considerable advantages of a market economy. Worker control has been, and is being, taken seriously by a number of thinkers. These range from John Stuart Mill, who saw it as the only hope for humanity's continued progress (1970: 133), to many socialists today who, having become disillusioned with central planning, see worker control as a way to achieve the advantages of both socialism and a market economy.

The main idea of worker control is simple: do away with the private ownership of corporations, partnerships, and other business establishments; turn them over instead to the workers themselves—all the workers, blue collar and white collar alike. Let the workers then manage the business democratically, or elect managers to do so, and share the profits. This way businesses will be oriented towards the interests of those who actually do the work, not those who merely own the capital. No longer then will ordinary workers get the worst of everything.

By 'worker control'—often referred to by other names such as 'workplace democracy' and 'economic democracy'—what is meant here is any economic system that features democratic control by workers over the establishment for which they work. (And note that, throughout, the term 'worker' refers not just to the blue-collar employees of an enterprise, but to anyone employed by the enterprise from the highest levels of management down. It has been suggested to me, incidentally, that the term 'employee control' might sound less radical and thus less threatening to traditionalists, but, for the sake of compatibility with others writing in this area, I shall retain the term 'worker control'.) By 'socialism' what is meant here is any economic system that features state ownership and control of the means of production. Given this terminology, a worker-control system need not necessarily be a socialist one; worker-control capitalism is a possibility also. By 'worker-control capitalism', what is meant here is any system that combines worker control with *private* ownership of the means of production. (Worker-control socialism and worker-control capitalism are, incidentally, sometimes referred to as two different versions of 'market' socialism. But since worker-control *capitalism* does not feature any state ownership of the means of production or central planning, to call it socialism of any sort is misleading.)

In this chapter, I shall critically examine worker control, both as a socialist and as a capitalist system. A version of worker-control socialism actually existed for a number of years in what was formerly Yugoslavia, with mixed results. But, in examining worker-control socialism, I shall not focus upon this version, with all its ambiguities and defects. Rather, as I did in examining libertarian capitalism and central-planning socialism, I shall focus upon an idealized version of worker-control socialism—namely, the version set out by David Schweickart in his challenging book: *Capitalism or Worker Control?* (1980). (For a survey of other recent worker-control proposals, see Christie 1984.) Then I shall try to sketch, in some detail, a plausible version of worker-control capitalism to compare with this socialist version. After considering the advantages and disadvantages of each I conclude that, contrary to what most advocates today of worker-control claim, the most promising version of worker control is the capitalist one. Finally I shall, from a moral perspective, compare worker-control capitalism with current capitalism. I conclude that, in the light of this comparison, worker-control capitalism deserves to be taken seriously.

4.2 WORKER-CONTROL SOCIALISM

The model of worker-control socialism that Schweickart proposes has three basic features: (1) each business enterprise is to be managed democratically by its workers; (2) most new investments throughout the economy are to be planned and chosen by the government; (3) aside from new investments, the economy is to rely entirely upon the free market, that is, upon supply and demand, rather than central planning (Schweickart 1980: 49–50). Let us look at each of these three features more closely.

The most striking and significant feature of worker-control socialism is, of course, worker control. What worker control entails is that, by law, each business enterprise within the economy is to be a wholly independent, autonomous unit that is run as a democracy with all the employees of the enterprise, from janitor to president, and only its employees, constituting the electorate, each having an equal vote in any company elections. In most smaller enterprises, ones with, say, ten or less workers, this will be a direct democracy, a democracy in which all major policy decisions—from what and how much to produce, to how net proceeds are to be distributed among workers—are to be decided by a direct vote of everyone. But with most larger enterprises, ones with over ten workers, some with thousands of workers, a direct democracy would be impractical. Direct elections on every matter of any importance would take up too much time and, in any case, would often involve technical matters about which the majority of workers would know little. Thus larger enterprises are to be run as a representative democracy. In these enterprises, the workers elect a management team to represent them, and it is this management team that will handle the day-to-day operation of the enterprise. But these elected representatives, in running the enterprise, must always keep the worker's best interests in mind because, after a certain term specified in the enterprise's constitution—say, every four years—they will have to run against an opposition slate of managers in another election, and if the workers have not been satisfied with the management team's performance, the team will not be re-elected. If they are not re-elected, they will then have to retreat to some less important (and lower paying) positions in the enterprise, or perhaps try to get hired, or elected, as managers at some other enterprise where they will, once again, have to withstand periodic worker elections.

How often elections are to be held, whether management will be subject to votes of non-confidence prior to regularly scheduled elections, whether any other matters such as major changes in the direction of the enterprise are to be decided by direct vote also, these are all things that are to be specified in the enterprise's constitution. This constitution, if it does not provide for a direct democracy, is to provide at the very least for periodic elections of managers and for a means of amending the constitution by a direct vote of all workers (with, perhaps, a favourable vote from at least two-thirds being required for approval). What else the enterprise's constitution provides is to be left up to each enterprise to decide for itself.

How exactly, it might now be asked, does worker control help alleviate the burdens that ordinary working people must bear? One way is by increasing worker autonomy, or self-determination, and thus by increasing worker 'self-realization'. No one enjoys being bossed around all the time. Satisfaction from work depends, in part, upon a worker's having some control over what he or she is doing. Without any control, without any autonomy, the worker is reduced to little more than an automaton, going through the motions of some assigned task, but gaining little satisfaction from it, and little realization of his or her potential. Worker control—an equal vote for everyone in specified company decisions—will help bring about more worker autonomy and self-realization.

We must be careful, however, not to exaggerate the gains in worker autonomy and self-realization. The gains will be greatest in certain very small enterprises in which direct democracy, and thus the full participation of all in the company's everyday decisions, is possible. But small enterprises in which this is possible will be the exception, not the rule. In the majority of enterprises, the only feasible form of worker control will be, as we saw, a representative democracy, where worker participation in decision-making is limited largely to periodic elections of managers. The gains in worker autonomy and self-realization provided by this more limited degree of participation will be more modest.

Yet even these more modest gains will constitute a significant improvement in worker autonomy over what exists in enterprises that are organized along the totalitarian lines prevalent today. A useful comparison is the gain in autonomy that people realize in going from a totalitarian *political* system to one that is a representative democracy.

But the gain in autonomy from a single vote is not the main advantage in going to worker control. Consider again the analogous case of going from a totalitarian political system to one that is a representative democracy. The main advantage of a democratic political system is not the autonomy a person gains from a single vote. The main advantage is that a representative democracy serves as a check upon the integrity of government officials, a check which prevents them from straying too far from the general welfare. Government officials in a representative democracy, knowing they must face the electorate periodically, are thereby forced to cater to the electorate's interests. Sometimes the prospect of upcoming elections causes government officials to cater to special interests rather than the general welfare. But, nevertheless, a representative democracy, in conjunction with a free and vigorous press, is likely, in the long run, to be more responsive to the general welfare than is a totalitarian system (see Sect. 3.6). Likewise, the main advantage of worker control is that the system serves as a check upon company officials, making it more likely that they will cater to the interests of workers. With worker control, in other words, company officials have more of an incentive to try to make working conditions as safe and pleasant as possible, to try to provide workers with a more favourable balance between work and leisure, to try to relieve the tedium of most ordinary jobs, and to try to allow workers some scope for exercising their creativity. In this way, more than any other, worker control helps relieve the burdens of ordinary working people.

And there is still another way in which worker control helps ordinary working people: by providing a somewhat more equal distribution of income and wealth throughout society. What each worker's share of the net proceeds is to be will be determined, if not by a direct vote, by elected managers. Notice, however, that since worker-control socialism is essentially a market system, prices, including salaries, are dependent upon the market, that is, upon supply and demand. Thus with worker control, just as with ordinary capitalism, those with scarce talents and abilities will generally be able to command a higher income than others. If those with scarce talents and abilities are not given an income commensurate with their talents and abilities, then (assuming they are not held in check by company loyalty) they will simply go elsewhere, to a company that is willing to pay them an income commensurate with their talents and abilities. This means, for example, that a company's managers, who presumably possess scarce managerial skills, will be able to command much higher pay than, say, the company's janitors. But in spite of these differences in income, there are at least two reasons why incomes in worker-control socialism will be more equally distributed than in current capitalism.

First, workplace elections will exert some pressure towards equality. The general idea is that, just as heads of state who are democratically elected tend to get less than those from totalitarian countries (compare, for example, the President of the US with the dictator of Iraq), likewise so will managers who are democratically elected. Opposing management candidates will be forced to compete with one another in promising not to grab too much of the company's net proceeds for themselves. And this competition is to be enhanced by a law requiring that the (real, not nominal) salaries of any elected company officers be made known to all company workers. Moreover, as may be the case in very small enterprises, if incomes are determined by a direct democratic vote of all workers, then the majority will likely be able to bring about still greater equality yet.

The second reason that incomes in worker-control socialism will be more equal than in current capitalism is that one very important source of inequality under current capitalism will be absent: namely, investment income, income merely from one's money, or capital that one's money has bought. As we saw earlier (Sect. 3.5), the opportunity that current capitalism affords for making money merely from money greatly favours the rich who have the money to do so; the rich, by making money from money, do indeed get richer, while the poor struggle merely not to fall further in debt. With worker-control socialism, on the other hand, private investments are not permitted; therefore, there is no opportunity for the rich to get even richer merely from their money. In short: to earn money in worker-control socialism, one must do it the old-fashioned way; one must actually *work*— that is, one must work personally rather simply having one's money work for one. And to the extent that the inequalities of income and wealth in current capitalism are in fact a result of the rich having their money work for them, income and wealth will be distributed more evenly in worker-control socialism.

But if, in worker-control socialism, investments are not handled privately, then how exactly are they to be handled? This brings us to the second major feature of worker-control socialism: governmental planning and control of investments. It works as follows. The government—which is to be democratically elected—will appoint a national investment board.

The members of this board are then to draw up a national investment plan, which specifies what the overall amount of investment is to be and how this amount is to be divided up among different types of investments. This plan might, for example, specify that 'A' per cent of the overall amount is to go to developing new energy sources, 'B' per cent to new housing, 'C' per cent to expanding the production of agricultural machinery, and so on. The funds for these investments are not, of course, to be supplied, directly, by private individuals, but are to be supplied by the government, which is to raise the funds through taxation. So-called investment banks—a national investment bank for investments of national scope, regional investment banks for investments of regional scope, and community investment banks for investments of community scope—are to be established for dispensing these investment funds to individuals. And the national investment plan, in addition to specifying investment priorities, is to specify also what per cent of the overall investment funds available is to be allocated to each of these banks.

The national investment plan is to be approved by the national legislature, and then, in the light of the national plan, planning boards for each region and community are to draw up their own plans, consistent with the national plan. And these plans are to be approved by regional and community legislatures. Although Schweickart does not explicitly say so, I assume each of these regional and community plans must be submitted to the national board for approval. Otherwise, adequate co-ordination of the regional and community plans with the national plan would appear to be unlikely—unless, that is, the national plan were written much more specifically than Schweickart suggests, thus leaving little or no leeway for innovation at the regional and community levels. In any case, after all these plans have been established, then individuals are to come to the investment banks with investment proposals. The banks are to accept from among these proposals those that are most promising and most in accordance with the particular investment plan the bank is charged with implementing.

Once the decisions have been made as to which individuals' proposals to accept, and at what level of funding, then the funds are to be distributed to these individuals, and they are then responsible for buying, on the open market, the capital goods—the land, factories, machinery, and so on—that they need for implementing their proposals. After they have bought the capital goods they need, and have advertised for, and hired the necessary workers, the enterprise is ready to proceed and, as it is to be self-supporting and wholly autonomous, it will proceed without governmental input or interference. Whether it succeeds or fails will therefore be a matter of supply and demand, just as it is for a traditional capitalist enterprise. But, of course, control of the enterprise and the profits from it are to be split among the workers themselves rather than among entrepreneurial owners of capital.

Schweickart stipulates that, with this version of worker-control social-ism, each enterprise, by law, must, from its profits, set aside enough depreciation funds each year to keep its factories and machinery in good working order and to replace them after their usefulness has expired. Each enterprise must also pay to the government, each year, a tax that is equal to a certain percentage (say, 10 per cent) of all the investment funds the enterprise has received from the government. The revenue from this tax then goes to support new investments. (This tax is the counterpart of what, in traditional capitalism, is the interest paid on borrowed funds.) Finally, if an enterprise goes bankrupt, or disbands for any other reason, then the proceeds from the sale of any remaining capital goods go, first, to pay off any creditors, and the remainder goes back to the government.

Since, with worker-control socialism, each enterprise is to have complete autonomy in everything other than investment decisions, it will determine its own methods of production, prices, income distribution, and so on. So, aside from investments, with worker-control socialism the market, not central planning, governs all aspects of the economy. This is the third major feature of this system. The significance of this feature is that, for the most part, worker-control socialism therefore shares with capitalism the motivational advantages of the market. Also, outside of the area of investments at least, it, like capitalism, avoids the problems that central planners have in gathering enough information to co-ordinate the economy successfully.

And, incidentally, not all of the government's investment funds are to go to individuals for starting new enterprises. Some funds are to go to existing enterprises for branching out in new directions. Moreover, existing enterprises are to be permitted to use certain of their own 'depreciation funds' for new investments even without government approval (Schweickart 1980: 53–4). This is an attempt to give existing firms some autonomy even with respect to investment decisions. But since the amount of depreciation funds to be used for new investments is, by law, limited, the degree of investment autonomy this provides is not extensive. Finally, if the various investment proposals put forth originally by individuals and existing firms do not add up to a right mix of investments for satisfying the government's overall investment plan, then the government is free to fill in the gaps by initiating new investments itself. But, of course, once a governmentally initiated investment project is under way, it is then to be turned over entirely to the workers, from which time on it is, like any other worker-controlled enterprise, entirely on its own.

What are the alleged advantages of government planning and control over investments, as Schweickart proposes? First, this system, as we have seen, eliminates one major source of inequality in wealth and income: the opportunity of making money merely from money or the ownership of capital goods. Secondly, government planning and control of investments will constitute a weapon that government can use in fighting unemployment and business recessions. The idea is that, with this control over investments, the government can direct them into those regions that, during hard times, need economic stimulation the most. The government can also direct investments into certain industries that appear to be especially crucial for healthy economic growth or for other legitimate reasons (such as, say, a viable national defence), investments which, for one reason or another, might not be forthcoming if left entirely to the market.

Schweickart argues that government planning and control of investments will also be more conducive to controlling negative externalities, such as pollution. In capitalist countries the government is, of course, capable of controlling externalities also, through various sorts of special legislation (see e.g. Blinder 1987: ch. 5). Schweickart, however, argues that, so as to maximize profit, adequate control of negative externalities is often not built into new investment projects under capitalism. The public then does not become aware of the need for controls until interests have already become vested—that is, until the offending establishments are already operational and it would then be more costly than ever to instil the necessary controls. So negative externalities, under capitalism, often remain uncontrolled. But since, with governmental planning of investments, the general welfare, not mere private profits, is the goal being sought, the appropriate externality controls are therefore more likely to be built into projects from the start. Schweickart thus concludes, and perhaps with some justification, that in controlling negative externalities, an economy in which the government controls investments has somewhat of an edge.

This then completes my sketch of the main features of worker-control socialism, as set out by David Schweickart. How desirable this system is remains now to be seen.

4.3 PROBLEMS WITH SOCIALIZED INVESTMENT

So far we have seen that worker-control socialism has some significant advantages over both central-planning socialism and current capitalism. But, as we shall see, it also has some serious disadvantages, or problems, which must be weighed against these advantages. Let me turn now to a set of problems that revolve around one particular feature of this system: government control and planning of investment or, in other words, 'socialized' investment.

4.3.1 The Élitism Problem

Earlier (Sect. 3.6) I argued that, just as the best way for the truth to emerge is through free competition among all ideas, likewise the best way for those goods and services that are most in people's interests to emerge is through a free-market competition among all products. (Consumer health and safety regulations, and the control of externalities and monopolies can, of course, be justified, but this does not affect the argument.) Assuming this argument is sound, it follows that a free market that caters to people's preferences is preferable to central planning that overrides these preferences. And, as we saw (Sect. 3.6), one reason why central planning overrides people's preferences is that central planners generally have inadequate information about what people's preferences are. But since worker-control socialism is essentially a market economy, it avoids this information problem of central planning. The members of the investment planning board will, with worker-control socialism, generally have as much information about people's preferences, as reflected in market prices, as does a well-informed capitalist entrepreneur. Thus these investment planners can, in principle, plan the economy's investments to reflect people's preferences as accurately as do the investments in traditional capitalism.

To what extent, however, will these planners in fact choose investments that reflect people's preferences? Might they not instead choose, for example, to invest in opera houses even though they know people prefer football stadiums? With all their power, it is unlikely that these planners will always be able to resist the temptation of trying, in their investment decisions, to 'improve' upon what people's preferences call for. And (assuming my earlier argument is sound) this sort of government meddling with people's free choices among products is no more appropriate than the censor's meddling with people's free expression of ideas. In short, with socialized investment, unwarranted élitism from central planners is a constant danger.

And the fact that the investment plans are to be reviewed by democratically elected legislative bodies does not eliminate the problem. Legislators may be able to tinker with complex interrelated plans such as these only to a limited extent. Moreover, the legislators themselves may not recognize the extent to which people's preferences have been inappropriately disregarded. Finally, the inappropriate neglect of these preferences might occur instead at the investment bank decision-making level, in deciding which particular investment projects to fund, and the decisions of these investment banks are not subject to legislative review.

4.3.2 The Tyranny of the Majority Problem

Related to the élitism problem is the tyranny of the majority problem. The élitism problem is that of governmental investment decisions neglecting *majority* preferences, while the tyranny of the majority problem is that of their neglecting *minority* preferences—preferences such as those for electric toothbrushes, caviar, and frisbee golf. Planners might well feel little pressure indeed to complicate matters, and expend valuable resources, merely for the sake of minority preferences such as these. And, of course, the will of the majority, as expressed through democratic elections, offers, in these cases, virtually no remedy at all.

4.3.3 The Red-Tape Problem

As Schweickart explains it, socialized investment is a very complex matter. First, an overall plan must be drawn up by a national investment board. For countries as large as the United States or Great Britain, this by itself would be an awesome task; all the parts of the plan must fit together properly, and in accordance (hopefully) with people's preferences. Then the plan must be approved by the national legislature. Hearings are to be held, and individuals and groups who oppose certain aspects of the plan are to be provided a forum to state their case. If the plan fails to be approved then, since all the parts of such a plan are interrelated, the whole thing must be more or less redone and resubmitted to the legislature. Thereafter, each of the country's regions and communities must draw up their own plans, compatible with the national plan, and each of these must be submitted to the corresponding regional or community legislature for hearings and approval. All those plans that fail to be approved must then be redone and go through the process all over again. And still another step might then be needed at this point, depending upon how much autonomy the regions and communities had been given in drawing up their own plans. If they had been given a fair amount of autonomy, then each of these regional and community plans might then have to be submitted to the national board so as to confirm their compatibility with the national plan, and with all the other regional and community plans. Otherwise a lot of inefficient duplication and omissions might creep into the overall picture. In any case, after all of this is finally accomplished, individuals, groups, and existing enterprises are to draw up investment proposals to be submitted to the national investment bank, and the various regional and community banks. These proposals not only will have to present a case for the proposed project's economic feasibility, but will also have to show how the project fits into the overall plan. Then the various investment banks evaluate each of the proposals, accept the ones that are most promising financially and fit best into the overall plan, and, finally, distribute the necessary funds to the winners.

This whole, complex process is problematic for more than one reason. First, given how slowly the machinery of democracy, and of governmental bureaucracy, usually moves, it might take a year, or even years, to get through the entire process, by which time the original plan might have become hopelessly outdated. And the costs of all of this bureaucracy and red tape might be far from insignificant. Finally, a process as slow and cumbersome as this is ill designed for adapting to changing conditions and unforeseen contingencies, and for taking advantage of opportunities while the time is still ripe.

Of course the red-tape problem could always be made less severe by cutting out certain steps in this cumbersome process, such as perhaps that of legislative approval. But then other problems, such as how to assume that the planners do not stray too far from people's preferences and the general welfare, would only become more severe.

4.3.4 The Intimidation Problem

Central-planning socialism, as explained earlier (Sect. 3.5), gives rise to the following intimidation problem. One risks forfeiting one's raises, one's promotions, indeed one's very job, by losing the good will of one's employer. And one risks losing the good will of one's employer by vigorously criticizing one's employer in public. But since, under central-planning socialism, the only employer throughout the whole economy is, in the end, the government, one thus risks forfeiting one's raises, promotions, and even one's very job by vigorously criticizing the government in public. Therefore, in central-planning socialism, people will naturally feel somewhat intimidated about doing this. Moreover, since a healthy democracy depends upon people's not being intimidated about vigorously criticizing the government in public, we may conclude therefore that a healthy democracy is, to that extent, incompatible with central-planning socialism. Since it calls into question the very possibility of ever having a healthy democracy, the intimidation problem of central-planning socialism is serious indeed.

Worker-control socialism does not, to be sure, give rise to exactly the same intimidation problem as central-planning socialism. With worker-control socialism the government is certainly not the only employer in the whole economy. Once businesses have been stocked with the necessary capital goods, they are free to make their own decisions however they may please, including all hiring, firing, and promotion decisions. Each worker-controlled enterprise thus constitutes a separate source of employment, just as does each enterprise in a capitalist economy.

But worker-control socialism nevertheless gives rise to an intimidation problem of its own. This is because there is at least one important respect in which the enterprises in this system can never be entirely free from government influence; they depend largely upon the government for the approval and funding of their capital expansion plans. Enterprises, according to Schweickart's scheme, have control over their own capital expansion plans only to the limited extent that they are allowed to use their 'depreciation' funds for this purpose. For any major capital expansion plans, they would need government approval and funds. It is this which gives rise to somewhat of an intimidation problem for worker-control socialism. Since an enterprise's capital expansion and thus growth depends largely upon government approval, and government approval might depend upon the enterprise's maintaining the government's good will, an enterprise may well, therefore, be reluctant to risk this good will through vigorously criticizing, or allowing its employees to criticize vigorously, the government. Even the media industry depends largely upon the government for growth money; so even those in this industry—newspaper reporters, editorial writers, newscasters, and so on—may be somewhat intimidated. Thus, although this problem will not, of course, be as serious as it is with full-scale central planning, with worker-control socialism there nevertheless may well be a problem, a problem with potentially serious ramifications for the effectiveness of political democracy and free speech within the country.

4.3.5 The Democracy Problem

The democracy problem comes in two versions. The first version of the problem stems from the reluctance of the typical electorate to sacrifice short-term interests for long-term interests. A politician who, in the name of long-term interests, calls for higher taxes or less consumer goods, is, typically, at a distinct disadvantage *vis-à-vis* an opponent who takes a shorter-term outlook. With worker-control socialism, the country's rate of savings and investment will be determined not largely by the market, as with capitalism, but by the electorate through democratic vote. And since the lower the rate of savings and investment, the more current consumption there will be, a democratically determined rate of savings and investment may well end up being irresponsibly low.

The other version of the democracy problem concerns special interests— that is, interests that are contrary to the general welfare and (not being protected by any rights either) are thus undeserving of governmental support. In a democracy, politicians, so as to get themselves elected, tend all too often to cater not only to the electorate's lack of long-term perspective, but also to any special interests that will get them more votes (see Sect. 3.6 above). In worker-control socialism, since the government is responsible for deciding which industries the enterprises are to get growth money and which not, the opportunities for catering to special interests are greater than in an economy where the government does not have this responsibility. If the government is responsible for handing out investment funds, then, regardless of what may be in the general welfare, politicians can all too easily say, 'Vote for me, and I will support more investment funds for your industry.' And since large, well-known enterprises can be expected to 'deliver' more votes, they may well get more funds than smaller, less well-known enterprises, no matter how much more promising the smaller, less well-known enterprises may happen to be.

And, as explained earlier (Sect. 3.6), the problem of special interests may be more serious in a socialist country than in a capitalist country for two additional reasons. First, an economic plan (whether for the whole economy or just for investments) cannot very well be voted upon by the legislature piece by piece; if it were, then the plan might well end up being incongruent. But if, therefore, the plan is instead voted upon as a whole, then this helps protect any of its components that cater to special interests. Thus such a plan may for this reason alone become a useful vehicle—one that does not exist in capitalist countries—for slipping special-interest provisions by the legislature unchallenged.

The other reason special interests may be more of a problem in a socialist economy is that what may be the major check upon governmental support of special interests—namely, vigorous, public criticism of the government—may not be as available in a socialist economy as in a capitalist one. This check may not be as available because of the intimidation problem.

4.3.6 The Incentive Problem

Finally, there is the very real possibility that, with worker-control socialism, people will not have as much of an incentive as they do in capitalism to start new enterprises and, in general, to innovate. It is true that, in starting new enterprises, people under worker-control socialism need not risk their own wealth, which may be an advantage in motivating people. This, however, may not be much of an advantage given that most investments, under capitalism, are financed by rich people who can afford the risk.

Of more significance are the disadvantages of worker-control socialism in motivating new enterprises and other innovations. First of all, under worker-control socialism one's proposal must more or less comply with the government's overall investment plan; otherwise it is probably doomed from the start. And even if it does comply, one may still fail to convince the appropriate government authorities that one's proposal is superior to competing proposals. To be sure, with Schweickart's scheme there will probably be more than just one investment bank to which one might submit one's proposal, thus increasing one's chances. And it may be no easy matter to convince funding sources of the merits of a proposal under capitalism either. But, nevertheless, it would appear that, with capitalism, the potential sources of funding—banks, stocks and bonds, independently wealthy people, existing firms receptive to new ideas, and so on—are more numerous than with worker-control socialism, and thus the probability of being funded will be greater. Next, in starting a new enterprise under worker-control socialism, one has no assurance that one will be able to stay in control; one might be voted out by the workers at the first opportunity. Finally, and perhaps most important, under worker-control socialism one generally cannot reap the financial rewards of the enterprise primarily oneself; one must share any such rewards among all the enterprise's workers. So with what may be a lower probability that one's new enterprise (or other innovation) will be put into effect, with the prospect of less control over it, and with less financial rewards from it even if it is put into effect, the incentives to start a new enterprise (or be otherwise innovative) may well, therefore, be less under worker-control socialism than under capitalism.

The problems set out above raise serious questions about the value of worker-control socialism. But they raise no questions about the value of *worker control*, for all these problems are applicable to the socialist version of worker control only. If worker control were possible without socialized investment, if, in other words, there were a viable form of worker-control *capitalism*, then none of these problems would exist. I think that there is indeed a viable form of worker-control capitalism, and, in the next section, I shall try to sketch what the main features of such a system might be. Thereafter, I shall compare the advantages and disadvantages of worker-control capitalism with both the socialist version of worker-control and traditional (i.e. current) capitalism.

4.4 WORKER-CONTROL CAPITALISM

The version of worker-control capitalism that I want to propose for consideration has three main features: (a) favourable credit terms for enterprises that have a worker-control format; (b) the specification and legislative recognition of what constitutes a worker-control format; and (c) a buy-out requirement. Let us consider each of these features in turn.

4.4.1 Favourable Credit Terms

The biggest problem in devising a viable form of worker-control capitalism, as it was with its socialist counterpart, is determining how new investments are to be funded—that is, new ventures and the expansion of old ones, the very lifeblood of an economy. We cannot, of course, look primarily to the state for investment funds, for that is the socialist, not capitalist, solution. Saul Estrin (1989) proposes that new investments be funded largely by means of 'holding companies', private companies whose sole business is investing in what they consider to be promising enterprises (cf. Miller 1989). But thus splitting the risks of investment and the risks of production between two different groups is, I suggest, inefficient and likely to cause more problems than it solves.

The solution I propose is, rather than having the state *give* workers the necessary funds for new investments as with worker-control socialism, instead have the state, or private banks within state guidelines, *loan* workers the necessary funds at favourable credit terms. This way all capital assets will remain privately owned, and capitalism will be preserved.

The main two alternative models for funding new investments through loans turn upon whether these loans are to be made by state or private banks. According to the first model, a state bank (with many branches) is to be established for administering these loans, which are to come from public funds. The key is that this bank is, before all else, to be a profit-making institution so that capital will be allocated efficiently and these loans will not be a drain upon taxpayers. That is to say, the state bank is to grant these loans on the basis of sound economic criteria only; in other words, only in those cases in which the loan can be expected to provide a reasonable return to the enterprise over the supply and demand equilibrium rate. This means, among other things, that interest rates are not to be subsidized.

The state-bank model, however, has several drawbacks. First, by placing the responsibility for making these loans in the hands of government, there is a danger that criteria other than merely economic ones will eventually be used for determining who is to get them, criteria that are political or ideological. The use of political rather than economic criteria would mean a less efficient allocation of funds and would jeopardize the state bank's profitability, leading, perhaps, to a substantial burden upon taxpayers. Secondly, if government funds are being used then, during a serious recession in which many of these enterprises failed, the entire burden would fall, ultimately, upon taxpayers.

The second model—the private-bank model—largely avoids these drawbacks, and is thus the model that shall be adopted here. According to this model, these loans to worker-controlled enterprises are to be administered only by private banks and other private lenders (e.g. life insurers and pension funds). Private banks and other lenders are to be enticed into granting these loans by having a certain percentage of the loan guaranteed by the government, without, however, the government having any input into which enterprises are to get the loans. I am assuming that the capital assets to be bought with the loan, or other business assets, will secure the remaining amount. The exact amount to be guaranteed by the government is a technical matter that we need not go into here; the general idea is to make the guaranteed amount, say 30 per cent, just large enough to unloosen private lenders' purse strings, but without unloosening them to the point of recklessness. With, say, only 30 per cent being guaranteed by the government, in a serious recession taxpayers will therefore have to bear only 30 per cent of any losses from bankruptcies, rather than the entire amount. Moreover, since the private lenders will remain at risk for the unguaranteed amount, say 70 per cent, they will be careful not to make these loans except in those cases where the risk is reasonably good. Clearly if it is private lenders, without government input, that decide who is to get these loans, we can be sure that only economic, not political criteria will be used. And to increase the chances of successful repayment more, the government is to grant the privilege of making these partially guaranteed loans—a new and profitable source of business for private lenders—only to the most well-established and reputable banks and other private lenders throughout the country, and any of them that suffer an unusual number of defaults from these loans is to have this privilege revoked. Finally, in return for the privilege of making these partially guaranteed loans, lenders are, of course, to agree to granting the worker-controlled enterprises those favourable terms that the government specifies. And enterprises receiving these loans must, for their part, agree to a worker-control format, which includes the requirement that adequate depreciation funds always be maintained so as to further protect creditors (see below).

But, then, of what, exactly, should these 'favourable terms' consist? I have already suggested that, for the sake of an efficient allocation of capital and the integrity of the system, these favourable terms probably should not consist of unusually low interest rates. Instead, the main feature of these favourable credit terms. as I see them, is that the rate of interest or, probably better, the amount of principal to be paid each year is to vary according to the enterprise's yearly, average revenue per worker.

Varying the principal to be paid works as follows. Yearly, average revenue per worker is to be determined by adding together interest due, depreciation, and other fixed expenses, but not, of course, any salaries, perks, or bonuses, then subtracting the resulting sum from gross revenue, and dividing the difference by the number of workers. In those years in which the enterprise's average revenue per worker falls substantially, the principal due that year is to fall also. Conversely, in those years in which average revenue per worker increases substantially, the principal due is to increase. This would mean that, in lean years, less, or perhaps even no principal would need to be paid on the loan; then, so as to compensate, greater than normal principal would be paid in fat years. Thus varying the principal to be paid will have the effect of extending or contracting the length of the loan. Lenders (in accordance with governmental guidelines) are to try to set the variable payment schedule so that, in the end, these extensions and contractions will, more or less, cancel one another out. It would be simplest, and probably most advantageous to worker-controlled enterprises, if, with each such extension or contraction, the interest rate were to remain fixed. If however, fixed rates along with these extensions and contractions were to leave banks and other lenders too vulnerable to unpredictable yearly changes, then perhaps it could be agreed that, with each such extension or contraction, the interest rate is to be adjusted so as to reflect current rates more accurately.

But the technical details of these 'variable payment' loans need not concern us here. The general idea is simply for worker-controlled enterprises to pay less than average in lean years, and more than average in fat years, so as to minimize their vulnerability to year-by-year fluctuations in income, thereby providing worker-controlled enterprises with much the same financial flexibility as enterprises funded largely through equity (i.e. through shares held by the general public upon which dividends either may or may not be paid, depending upon the enterprise's current financial condition).

If, as time goes by, additional capital funds are needed, a worker-controlled enterprise will be free to get these funds in virtually any way it can, such as through additional loans, either with the favourable terms outlined above or without. Another possibility is the sale of bonds to the general public. See, in particular, Roger McCain's interesting suggestion (1977) that equity-like 'risk-participation' bonds be issued by worker-controlled enterprises. McCain argues that, with the aid of both ordinary and risk-participation bonds, a worker-controlled enterprise whose object-ive is maximizing profit per worker would 'attain the same allocation of resources as would a capitalist corporation, under comparable circum-stances and informationally efficient markets' (1977: 382). A worker-controlled enterprise might even raise funds by offering non-voting shares in its ownership to the general public (Jay 1980). In short, a worker-controlled enterprise has any number of alternative fund-raising sources available to it.

Indeed, the only fund-raising source not available to a worker-controlled enterprise is that of offering *voting* shares to the general public. Why not, it might be asked, allow voting shares to be sold to the general public, provided the general public's voting interest remained a minority one? If this were allowed, then wealthy managers of worker-controlled enterprises could buy the shares themselves (or have their secret representatives do so), which, even though they would have just a minority voting interest, might nevertheless be enough to give them complete control over the enterprise. This, of course, would defeat the purpose of worker control.

4.4.2 The 'Worker-Control' Format

In return for receiving these favourable credit terms, an enterprise must agree to be worker controlled and remain so (even after any such loans have been repaid). An enterprise is 'worker controlled' if and only if it has adopted a worker-control format—a format that is to be accorded full legislative, or legal, recognition and protection, just as is a corporate format today. The main defining characteristics of a worker-control format are as follows.

4.4.2.1 Workplace Democracy

Any enterprise adopting a worker-control format must draw up a constitution for itself that will establish either a direct or representative democracy within the enterprise, with all permanent workers, from managers to janitors, each having an equal vote. This constitution is to be subject to amendment at any time by, say, a two-thirds vote of the workers.

If, as would be the case with any larger enterprise, the constitution establishes a representative democracy, then the next constitutional matter to be decided is what positions are to be elected ones. At the minimum, the position of chief executive officer (CEO) is to be an elected one. But, in the interests of (election) efficiency and, perhaps, management continuity, it may well be the case that few, if any, other positions will be designated elected ones. Those positions not designated elected ones are, of course, to be filled the traditional way, through appointment on the basis of merit. (Notice that electing just the CEO, or CEO and a chief assistant, should be sufficient to cause the entire management 'team' to be responsive to worker interests—compare the election within the executive branch of just the president and vice-president in a political democracy.)

4.4.2.2 The Buy-In Requirement

Each permanent (i.e. non-temporary) worker that goes to work for an enterprise operating under a worker-control format must immediately buy, from the enterprise, a share of ownership in it that is equal in voting rights to the share owned by every other permanent worker. The price to be paid for this share is to be equivalent to the current value of the enterprise divided by the number of permanent workers there are (including the worker who is buying the share). This share is to be paid for by the worker through the method of 'deferred' payment explained in Section 4.4.2.5 below.

4.4.2.3 The Pay-off Requirement

Whenever a worker, for any reason, leaves an enterprise operating under a worker-control format, the enterprise is to pay the worker (or the worker's estate) an amount equal exactly to the current value of the worker's share, minus any deferred payment owed (see below). The current value of a worker's share is equal to the current value of the enterprise divided by the number of workers there are (including the one who is leaving). If, short of selling assets it prefers to keep, the enterprise does not, at the moment, have the cash to pay off the worker, it may borrow the money at the favourable terms described above. A worker is not to be permitted to divest himself of his share except by leaving the enterprise. (In certain cases of hardship, however, the enterprise, or a bank, should perhaps be allowed to grant the worker a loan using his or her share as collateral, although it would be collateral that could not be possessed until the worker left the enterprise.)

The pay-off requirement is important. Without the pay-off requirement, older workers would generally have little interest in the enterprise ever diverting any of its revenue away from current salaries to new investment. This is because older workers generally cannot expect to be with the enterprise long enough for any new investment to be worthwhile for them in terms of higher future salaries. Thus, for them, any revenue devoted to new investment would be largely wasted. Accordingly, these workers would generally oppose any new investment, which, in turn, would tend to result in an overall amount of investment that was suboptimal. This is sometimes referred to as the 'time-horizon' problem. With the pay-off requirement, however, no matter how soon a worker expects to leave the enterprise he will, upon leaving, tend to get a fair share of any new investment back in the form of this pay-off (which will be higher than otherwise because of the increase in the enterprise's value attributable to the investment). Thus, with the pay-off requirement, this particular time-horizon problem is largely avoided. (But see Sect. 4.6 below for more on time horizons.)

4.4.2.4 Limited Liability

If an enterprise operating under a worker-control format goes bankrupt, creditors are not, by law, to be able to reach the personal assets of the workers, just as they are not able to reach the personal assets of the shareholders of an enterprise operating under today's corporate format. In other words, legislative recognition of the worker-control format is to include 'limited liability' for workers.

In return for limited liability, however, the workers must submit to one condition. They must always maintain sufficient depreciation funds for the maintenance and replacement of their existing capital. Replacement assets should be of at least equal value, though not necessarily of identical type or function. Enterprises will, of course, be allowed to cut back their operations by not replacing used-up capital assets, or by selling them off. But any depreciation funds that would have been used to replace these assets, or whatever funds the enterprise may get from the sale of these assets, must first go to reduce the enterprise's debt; only after all debt has been eliminated may any such funds then be distributed to the workers (either outright or in the form of higher than usual wages). If a worker-controlled enterprise which has been granted limited liability on the condition that it maintain adequate depreciation funds fails to meet this condition, then the workers are to lose their limited liability, thus becoming liable personally for any company debts, with the extent of their liability being proportional to their relative salaries within the enterprise.

4.4.2.5 Deferred Payments

We come, finally, to the question of how workers are to pay for the share that they are obligated to buy from a worker-controlled enterprise upon joining it. Notice, first of all, that this should be no problem for workers joining a newly established enterprise prior to the start of its operations. Typically, a newly established worker-controlled enterprise will be financed entirely through loans and, therefore, at the beginning of its existence its debts will equal its assets, which means that, at the beginning of its existence, its value will be nothing. If the enterprise's value is nothing then, of course, the value of a share in the enterprise is nothing, and thus the initial workers will be able to acquire their share at no cost.

But as the enterprise, over the years, pays off its initial debt and, in general, prospers, the value of a share should grow accordingly. Thus the time may come when the cost to a new worker of a share in the enterprise will be substantial, which gives rise to several problems. First, requiring a new worker to pay what may be a substantial amount for a share, while the initial workers got their shares for nothing raises questions of fairness. Second, if new workers in worker-controlled enterprises had to either pay a lump sum buy-in price, or have their salaries reduced each month so as to pay for this share, then worker-controlled enterprises would have difficulty competing with traditional enterprises for new workers since, with traditional enterprises, new workers would be under no such financial constraints. Third, the value of a share in long-established worker-controlled enterprises that had done exceptionally well throughout the years might have reached hundreds of thousands of dollars, far beyond what any ordinary worker would be able to pay. Such firms would then be unable to hire any needed workers at all.

All of these problems are avoided by putting a new worker on a par with original workers, ones present at the beginning of the enterprise's existence who, as we have seen, typically acquire their shares at no cost. A new worker can be put on a par with original workers by determining the value of his share as follows. Whenever the new worker leaves the enterprise, instead of determining the value of his share, and thus pay-off, by dividing the relevant number of workers into the *overall* value of the enterprise, we determine it by dividing the relevant number of workers into the *increase* in the enterprise's value from the time the new worker began to the present. Accordingly, if, on the very day the new worker began he were to quit, he would get no pay-off at all, since, presumably, there would have been no increase in the enterprise's value during this single day, and thus his share would be worth nothing. This then puts the new worker on a par with original workers whose shares, at the time they began, were worth nothing as well.

Consider, for example, an enterprise with ten original workers, each of whose share at the beginning was worth nothing (since, at the beginning, the enterprise's liabilities equalled its assets). Say that after ten years, during which time the value of the enterprise has grown to $100,000, the original workers decide to expand by hiring ten new workers. At that time, even though each new worker is immediately to have full voting rights, the value of his or her share will be nothing. But say that, after still another ten years during which time the value of the enterprise has increased to $280,000, one original and one new worker decide to quit and receive their pay-off. The value of the new worker's share, and thus his pay-off, will then be $9,000, calculated by dividing the increase in the value of the enterprise since the new worker began ($180,000) by number of workers (20). The value of the original worker's share, and thus his pay-off, will be $19,000, calculated by adding the value of his share just prior to taking on the new workers ($10,000) to the increase in the value of every worker's share since then ($9,000).

If, on the other hand, the value of the enterprise had not increased to $280,000, but instead had decreased to $80,000, then the value of the new worker's share, and pay-off, would be nothing (no increase in value divided by 20 equals nothing), while the value of the original worker's share, and pay-off, would be $8,000 ($80,000 divided among the ten original workers). Finally, if, after still another ten years, the enterprise had a negative value of, say, minus $80,000, then the value of both a new and old worker's share, and their pay-off, would be nothing, although, because of their limited liability, the personal assets of neither could be reached to pay the enterprise's debts (provided that adequate depreciation funds had been maintained).

Determining these share values was simple enough. But, it may be objected, the examples were artificially contrived to make them simple; normally; with many people being hired and leaving the enterprise at many different times throughout the years, these calculations will get rather complex. Perhaps, but hardly so complex as to require high-level mathematics or stretch the capabilities of today's computers. Moreover, a simpler way to conceptualize how the value of a new worker's share is to be determined, a way that is mathematically equivalent to the way suggested above, is as follows. Let us say that any new worker must 'buy' a full share in whatever worker-controlled enterprise at which he or she starts working, and that the price of this share shall be equal to the current, overall value of the enterprise divided by the number of workers (including the new worker). But—and this is the crucial part—the new worker need not pay for this share until he or she leaves the enterprise, at which time the full purchase price will simply be subtracted from the current value of the worker's share, thereby reducing his or her pay-off. Thus the new worker is to be thought of as owning a share equal in value to that of anyone else in the enterprise at the moment he or she begins working full time. But payment for this share is to be deferred until such time as the worker leaves the enterprise. We may thus speak of this as buying into an enterprise through the method of 'deferred payment'. And if whenever the new worker leaves the enterprise the current value of his or her share happens to be less than the buy-in price he or she 'owes', then (so as to preserve limited liability and maintain his or her parity with the original workers) the new worker need not pay the difference. He or she will simply receive no pay-off.

Consider, again, the example above. According to this way of conceptualizing matters, each of the ten new workers must, upon starting full-time at the enterprise, buy a full share, which will cost each worker $5,000 ($100,000, the current value of the enterprise, divided by 20, the number of workers including the new ones). After ten years, during which time the value of the firm gradually increases to $280,000, each worker's share then equals $14,000 ($280,000 divided by 20). If one new and one original worker then decide to quit and receive their pay-off, the new worker's pay-off will be $9,000 ($14,000 minus his deferred buy-in payment of $5,000) and the original worker's pay-off will be $19,000 ($14,000, plus his one-tenth share of the $50,000 in deferred payments that the original workers get from the ten new workers). But if the value of the enterprise had, during this time, decreased to $80,000, then the value of each worker's share would be $4,000 instead. The new worker's pay-off would then be nothing because he owes a $5,000 deferred payment. Only $4,000 of this amount would actually be paid, however, since (as explained above) he is not to be liable for the difference. The original worker's pay-off would equal $8,000 ($4,000, plus his one-tenth share of the (now) $40,000 in deferred payments that the original workers are to get from the new workers). So by conceptualizing the financial rights and duties of new workers in terms of 'buy-ins' and 'deferred payments', we reach the same results as before. This latter way of conceptualizing matters is the one to be adopted here. The former, equivalent way of conceptualizing matters was introduced largely because it helps clarify an important point; the shares new workers are to buy are not really 'purchases' in the traditional sense and do not increase the overall value of the enterprise. Rather, they are, in effect, 'disclaimers' of that which rightfully belongs to the old workers, namely, the enterprise's value prior to the new workers. Since these purchases are really disclaimers, to require any interest to be paid on the deferred 'purchase' price would be inappropriate.

This completes my outline of the 'worker-control format'. The main features of this format are, once again, workplace democracy, the buy-in requirement, limited liability, the pay-off requirement, and, finally, deferred payments.

In opposition to making favourable credit terms available to enterprises that adopt a worker-control format—terms designed to encourage enterprises to adopt this format—it might be argued that if these enterprises would not exist without these favourable credit terms, then they should not exist at all. Worker-controlled enterprises are not prohibited in capitalist countries today. Therefore, it might be argued, if they do not exist already, then this must mean that workers have freely chosen not to establish them. And if workers have freely chosen not to establish them, this must mean they do not want them. Finally, if workers themselves do not want worker-controlled enterprises, then there is no good reason for promoting them with favourable credit terms established by the government (cf. Nozick 1974: 250–3; Williamson 1980: 33–5).

Such an argument would not be convincing. The reason why few worker-controlled enterprises exist in capitalist countries today has little to do with what workers may or may not want. The reason they do not exist is that any individuals with sufficient initiative, knowledge, financial backing, and so on, to establish a new enterprise in capitalist countries today—whether these individuals be 'workers' or not—have a choice. They can choose to establish a traditional enterprise in which they, themselves, can be permanent bosses and keep the profits largely for themselves. Or, alternatively, they can choose to establish a worker-controlled enterprise in which they can be only conditional bosses, without keeping the profits largely for themselves. Obviously most enterpreneurs will choose the traditional enterprise, and this no doubt is why so few worker-controlled enterprises exist in capitalist countries today. But it should be equally obvious that this hardly means that the majority of workers prefer traditional enterprises, or that traditional enterprises are more in the general welfare.

Compare this with the colonization of the New World by adventurers several centuries ago. Consider, in particular, those adventurers with political ambitions; those, in other words, who wanted, personally, to govern the lands they would be colonizing in the name of whatever king was funding their expeditions. Say this king gave them a choice between either establishing a totalitarian colony in which they would be the permanent ruler with unlimited opportunities to enrich themselves or establishing a democratic colony in which they would have to face periodic elections, and even if elected would have only limited opportunities to enrich themselves. A very few exceptionally high-minded adventurers might choose to establish a democratic colony, but we can be certain that the vast majority with political ambitions would instead choose to establish totalitarian colonies, with themselves as ruler. But, clearly, just because these adventurers would choose to establish totalitarian colonies, this does not mean that totalitarian colonies, not democratic ones, ought to exist. Likewise, just because most entrepreneurs today (whether from the 'working' class or not) choose to establish traditional enterprises, this does not necessarily mean that traditional enterprises, not worker-controlled ones, ought to exist. (Cf. Walzer 1983: 295–303; for further reasons why worker control, even if morally and economically superior to traditional capitalism, might nevertheless not be able to compete under capitalist 'rules of the game', see Putterman 1982.)

4.4.3 The Buy-Out Requirement

So far then I have sketched two important features of worker-control capitalism: legislative recognition of the worker-control format and favourable credit terms for enterprises willing to adopt this format. These features should encourage the creation of some worker-controlled enterprises. But would they by themselves be enough to transform ordinary capitalism into what might justifiably be called 'worker control' capitalism? Probably not. To bring about such a transformation fully would require an additional feature to assure that all enterprises in the current economy would, sooner or later, become worker-controlled ones. But what should this feature be?

One possible step in the direction of a full transformation would be to require all new ventures henceforth to be worker-controlled ones only. The problem with such a requirement is that this might well create an incentive problem similar to the one pointed out earlier for worker-control socialism. If all new ventures had to be worker controlled then, just as with worker-control socialism, prospective entrepreneurs would not be able to keep capital returns largely for themselves, but would have to share these returns with workers. And they could not even be sure of maintaining their initial position of leadership in the new enterprise, for they could always be voted out by the workers. All of this might well lessen people's incentives to initiate new ventures, thereby hampering economic growth.

Besides, as Robert Nozick suggests, there is indeed something disturbing about a prohibition upon 'capitalistic acts between consenting adults' (Nozick 1974: 163). For the sake of personal freedom as well as maintaining incentives, people probably should be allowed to enter into whatever business arrangements they so please, provided only that the legitimate interests of third parties are not thereby compromised, and that all parties to the arrangements have indeed freely consented to them.

But if, on the other hand, all new ventures are not required to be worker-controlled ones then, as we have seen, entrepreneurs, for reasons of personal gain and control, will naturally tend to establish traditional enterprises instead. The availability of favourable credit terms for those willing to choose a worker-control format should counter this natural tendency somewhat. But, I suspect, many or most new ventures would be traditional ones nevertheless. In short, the problem is this: if new ventures with traditional formats were prohibited, then it is questionable whether this would allow for adequate incentives and freedom, but if they were not prohibited, then it is questionable whether the system would ever come close to being a truly worker-control one.

The solution I propose to this problem is to place no restrictions whatever upon the establishment of new ventures with traditional formats, but to impose a worker 'buy-out' requirement upon all enterprises. According to this buy-one requirement, any existing, traditional enterprise can remain in the hands of its current owners as long as these owners please, but whenever the owners either die or, alternatively, decide to sell, then the enterprise is to be bought by the workers and the workers only, and, from that point on, is to operate as a worker-controlled enterprise. Whatever amount the workers pay is to go entirely to the former owner, or to his or her estate. We may assume that, typically, this purchase will be financed by taking advantage of the favourable credit terms available to worker-controlled enterprises. This purchase by the workers will be similar to the purchase of a house in which the purchase price is paid by means of a long-term loan, except that no down payment will be required. The purchase price is to be equal to the current market value of the enterprise, as determined by special, government-appointed value assessors who are to determine the value of the enterprise in a way similar to how the value of a business is determined today for estate tax purposes (with each side having a right to appeal). And, so as to protect the workers' limited liability (see above), it is actually the enterprise itself, not the workers, that is to be directly liable for the repayment of this loan. With the purchase, however, each of the current workers thereby becomes the owner of a full, voting share in the enterprise—a share somewhat similar to the shares that people today own in corporations (although these shares, being tied as they are to employment in the enterprise, are not to be bought and sold on the market). But this share will cost the workers nothing initially since—given that the enterprise will be starting operations under its new worker-control format with a debt precisely equal to its value—each share will, at that moment, be worth nothing. Yet as the enterprise, over the years, pays off the debt, the value of each worker's share will, of course, grow accordingly. And, of course, having been purchased by the workers, the enterprise is henceforth to operate under a worker-control format, as described above.

But how does this buy-one requirement work in cases where the original ownership of the enterprise is split, either among partners or a number of shareholders? In these cases, the buy-out by the workers will proceed piecemeal, as each of the partners, or shareholders, dies or sells out. With the departure of each partner in a partnership, the workers are to purchase his or her interest so that, for the time being, the workers, as a group, will become equal 'partners' with the remaining partners. And in the case of a corporation, where the original owners take the form of shareholders, the workers are to acquire the shares bit by bit, as each shareholder dies or decides to sell. Of course whenever ownership thus passes gradually, piece by piece, to the workers, accountants must make sure that any amounts the enterprise may thereby end up paying to the bank on behalf of the workers come entirely from the workers' share or salary, so that the remaining, original owners do not end up subsidizing the workers' purchases.

The requirement that existing shares be sold only to workers may, for the most part, put an end to the stock market as we know it, but I, for one, will shed no tears. Only the purchase of *newly issued* shares directly from corporations serves the significant social purpose of supplying funds for new investment. The vast majority of sales on the stock market are merely transactions from one shareholder to another—often no more than a form of gambling for the rich. How much more productive might the economy be if the brilliant minds of Wall Street—the brokers, the advisers, the wheeler-dealers of all sorts—were free to contribute to the economy in some other way? And no longer would disgusting stock manipulators, such as Boesky and Milken, make hundreds of millions of dollars doing what has less social value than collecting garbage. Granted, in order for people to be willing to buy newly issued shares, they must know that there will be buyers available whenever they want to sell them. But these buyers need not necessarily be other (traditional) shareholders; they can be workers instead.

There are, to be sure, a number of troublesome details with the worker buy-out scheme outlined here that would have to be worked out. One potential problem is that of corporations selling newly issued shares for risky ventures to sham buyers, who immediately sell-out to the workers, thus, in effect, forcing the workers to bear the risk. But a number of possible remedies for this problem are available such as that of qualifying the mandatory buy-out requirement so that, for newly issued shares, it does not take effect until, say, five years from the date of the share's original sale; with any resale to individuals other than the workers being therefore permitted prior to the expiration of these five years. After five years, the new investment's potential for success should be much more clear, the purchase price of the share will have altered accordingly, and the risk greatly diminished. Another troublesome detail might be that of coming up with some provision for preventing corporations from delaying indefinitely the transition to worker control by the sale (or gift) of newly issued shares whenever the workers threatened to gain majority control. With small, family corporations, for example, the original shareholders could perhaps give enough newly issued shares to their children, grandchildren, or even great grandchildren, to keep the family in control of the corporation long after the original shares had passed on to the workers. The lifetime inheritance quota proposed in Chapter 6, however, would go far in blocking this sort of manoeuvre. But I shall not pursue details such as these here; I assume they can, with a little ingenuity, be worked out satisfactorily.

A worker who does not want to participate in a buy-out can, of course, always quit. But if a worker remains with the firm, he or she must participate. And participating in a buy-out may, to be sure, result in somewhat of a cut in pay so that the enterprise can meet the payments due on the buy-out loan. And a cut in pay for those purposes is, of course, equivalent to an 'investment' in the future of the enterprise. Moreover, as critics are fond of pointing out, such an investment does not have the safety of 'portfolio diversification'.

Yet, before feeling too sorry for workers who either must quit or invest in their enterprise's future, let us keep the following in mind. First, governmental guidelines will call for banks to spread the payments on the buy-out loan over enough years to prevent the burden upon workers from being excessive. Moreover, if the value of an enterprise is great enough to necessitate spreading the payments out over a number of years, then it is probably a well-established, healthy enterprise that is likely to continue to prosper. Such enterprises are relatively safe investments that are likely to yield decent returns over the years and create for each worker a substantial 'nest egg' for when he or she retires (over and above any ordinary pension funds). Indeed, the profits from such enterprises may well be enough even to cover the monthly payments due on the buy-out loan, in which case the workers' investment will end up being self-financing. If, on the other hand, the enterprise to be bought by the workers has not been doing particularly well and has debts of an amount that approach its assets, then, to be sure, the risk will be a poor one. But this should not be a serious problem for the workers either, for in that case the enterprise will be worth little or nothing, and being worth little or nothing, the cost of the workers' investment will be little or nothing as well. Thus, if the enterprise did fail, the workers would be little or no worse off financially than had they not invested. Remember, they, like traditional shareholders, are protected by limited liability (see Sect. 4.4.2.4 above). And if the enterprise is unattractive enough, then, of course, the workers may, on the advice of their bank, choose not to invest in it at all, in which case the enterprise will simply cease to exist, and its assets will be sold to satisfy the claims of creditors. Therefore, with the buy-out requirement, workers, generally, should come out rather well; often their investment will be very low risk, or it will cost them very little if anything, sometimes even both. And in all cases they finally gain control over their own destinies.

To illustrate how worker-control capitalism with this buy-out require-ment works, consider the following case. Jones, an enterpreneur, establishes a small shoe manufacturing business that employs 100 people. At Jones's death, the business is worth $5,000,000, and Jones was making a yearly profit of $500,000. The workers are, at this point, to purchase the business, with the proceeds going to Jones's estate. The workers will finance the purchase through, say, a private bank, with terms, perhaps, something like the following: no down payment, with the principal to be paid back over 20 years, at 10 per cent interest. These terms may also provide for a partial government guarantee and variable principal payments as explained above, but we may disregard any such details here. With these terms, the business (being the legal debtor) will be repaying the bank (principal plus interest) at a rate of roughly $580,000 a year. But since the profit—that is, the amount remaining after expenses, including all 'regular' salaries—is $500,000 per year, the workers will end up sacrificing very little indeed in salary (only $800 per worker, on the average). Moreover, as the amount the business has paid back on the loan each year increases, so also, to a corresponding extent, does the value of each worker's share, which the worker will be paid off upon leaving. Let us say, for example, that the business prospers and that around fourteen years later its overall value has increased to $10,000,000 ($12,500,000 minus the $2,500,000 still owed to the bank). The value of each worker's share will then be $10,000,000 divided by the number of workers, 100, which equals $100,000, and each worker, having acquired his share at no cost in the transition to worker control, will therefore owe no deferred payment for it. So if, at that time, a worker wishes to leave, his pay-off will be $100,000. And not only will workers be getting an investment for the future, they will be getting a management with greater incentive than before to please them, which may in turn mean better working conditions, more challenging work tasks, and so on. They will also be getting greater job security. And they will be getting autonomy.

But say that the yearly profits were, instead, only $100,000 and prospects for improvements in the future were not exceptional. With the traditional capitalism of today, it would be unlikely that any buyers willing to keep the business intact could be found. This is because, relative to interest rates today, a $100,000 yearly return on a $5,000,000 investment is not much. What would be the case, however, with worker-control capitalism? For much the same reason, a bank advising workers on the buy-out would likely recommend that they not pursue the purchase. Some attempt might thereafter be made to see if any larger worker-controlled enterprise would be willing to buy out the business and keep it going as a subdivision. But, short of that, the business would likely then cease operations (just as it probably would with traditional capitalism as well), and its assets would be sold and put to some more profitable use.

But, if the workers become owners, what about the risk they take of substantial fluctuations in pay from year to year? The answer is fourfold. First, by gaining control over their own destinies, their risk of being laid off merely for financial reasons becomes less (see Sect. 4.6). To the extent, therefore, that workers do in fact increase their risk of sudden fluctuations in pay by going to worker control, this may well be more than compensated for by the decrease in their risk of being laid off. Secondly, as Louis Putterman points out (1984), there are, in any case, ways to diminish any risks of sudden fluctuations in pay. These include the investment of portions of current revenue in diversified portfolios of bonds from other enterprises, or the participation of workers who are particularly risk adverse in insurance funds especially designed to ameliorate such risks. Thirdly, their risk of sudden pay fluctuations is diminished to the extent that the favourable repayment terms of their buy-out loan include 'variable' terms (see Sect. 4.4.1 above). Fourthly, such risks are also diminished by any governmental redistributive measures, such as those proposed in the next chapter.

More generally, with worker-control capitalism serious investments will no longer be largely the prerogative of the very rich, as they are with ordinary capitalism; instead, serious investments will become spread much more evenly throughout society, among numerous workers. The overall effect of this can only be that income and wealth will be distributed more equally.

So let us indeed not feel too sorry for the workers; in going from a traditional format to worker control, they will hardly be any worse off than before and will normally be a lot better off. They have little to lose and much to gain. (I often, incidentally, hear the argument that most workers simply do not want the responsibility of ownership and voting; that is, they do not want autonomy. I cannot help but be reminded by this of the argument, prevalent in the South prior to the abolishment of slavery, that most slaves simply do not want the responsibility of freedom; they are perfectly happy just as they are.)

One final point about the buy-out requirement is worth mentioning: its success in bringing about a *peaceful* transition throughout the economy. The most obvious and crude way to bring about a transition from traditional to worker-controlled enterprises throughout the economy would be for the government simply to force existing enterprises to be turned over, immediately and without compensation, to workers. This way, however, is morally unacceptable. It would be unjust to force capitalist owners simply to turn over their enterprises after the sacrifices they may have made in bringing about these enterprises, especially since they would have made these sacrifices on the understanding that they would be able to enjoy the rewards. Forcing all capitalist owners, immediately, to sell their enterprises to the workers would be better, for then at least the owners would get monetary compensation. Nevertheless, this would be to force them to give up their life's work at a time when many might prefer not to do so. The social upheaval and resentment that any such sudden transition would cause might well be enough to assure the failure of worker-control capitalism from the start.

But notice how nicely the buy-out requirement outlined here accomplishes the transition from traditional enterprises to worker-controlled ones. A law establishing worker-control capitalism by means of this requirement would, at the moment it took effect, leave everything exactly as it was before. The transition from traditional firms to worker-controlled ones would then proceed only gradually, bit by bit, as traditional capitalist owners either died, or voluntarily decided to sell. Coercion would be at a minimum, and all former owners, or their estates, would be fully compensated.

It might be objected, however, that the buy-out requirement does prevent the owners of businesses from leaving them to their children so their children can 'carry on in their footsteps'. Strictly speaking, this is true. But the interests of these children would not be compromised much. While owners were still alive and in control, they could place their children in key management positions and, if they did a good job, then, after workers took control, the children would likely be retained. If they did not do a good job, then it is an advantage, not a disadvantage of worker control that they might not be retained. Anyway, as long as the institution of inheritance were to be retained (see Ch. 6), then, even though the children of an owner could not inherit the business itself, they would nevertheless inherit the value of the business in cash. With this cash they could even, if they really so desired, start an identical business somewhere else. All in all, the buy-out requirement does indeed manage to bring about a transition to worker control with minimal interference in people's free choices.

We have before us now, in outline, a system of worker-control capitalism. Let us now see how the merits of this system compare with those of its socialist counterpart.

4.5 COMPARED WITH WORKER-CONTROL SOCIALISM

The main idea of worker-control capitalism is to realize the considerable advantages of worker control, while avoiding the problems inherent in socialized investment. I shall argue that the system of worker-control capitalism outlined here succeeds largely in doing this.

As argued earlier, the problems inherent in socialized investment include those of élitism, tyranny of the majority, red tape, intimidation, democracy, and incentive. Worker-control capitalism, as opposed to worker-control socialism, does not feature any government planning or control of investments. In fact, the version of worker-control capitalism outlined here, although a significant departure from capitalism as we know it today, is nevertheless about as pure a form of capitalism as can be found anywhere, since it features no government ownership of any capital goods, and no government planning of any aspect of the economy. For this reason, worker-control capitalism obviously avoids the élitism, tyranny of the majority, intimidation, democracy, and incentive problems of worker-control socialism.

Worker-control capitalism will also avoid the red-tape problem of worker-control socialism. But any gains from avoiding this problem will be offset somewhat by a different red-tape problem to which worker-control capitalism gives rise. The viability of worker-control capitalism depends upon there being accurate periodic assessments of the current value of worker-controlled enterprises. An assessment of current value is needed so as to arrive at a correct purchase price for purposes of applying the buy-out requirement. Thereafter, the current value of a worker-controlled enterprise must be determined at regular intervals for purposes of paying off departing workers, and arriving at the buy-in price for new workers. These intervals could perhaps be no more than yearly, in which case the exact value of departing worker's pay-off might not be known for sure until the end of the year in which he left, while a new worker's buy-in price might be determined by whatever the enterprise's value was at its last yearly assessment.

But even if, for the sake of minimizing administrative costs, assessments were not required more often than yearly, there still would remain the problem of how, exactly, these assessments are to be made. Determining the value of the enterprise, and thus the value of a share in it, would be no problem if the worker's shares were being sold on the market but, of course, they are not. An enterprise's current value will thus have to be determined from such facts as the current market value of its capital assets, the amount of its current debts and credits, its yearly revenue from ordinary operations, its future prospects, and so on. Determining current value will not, of course, be an 'exact science'. But it can, I think, be done within at least a tolerable margin of error, just as it has always been done for businesses that have not issued shares which are being sold on the market. Financial analysts will have to work out, as best they can, some formula for balancing these various factors and coming up with an appropriate figure. Perhaps in many cases the necessary calculations could be carried out by the enterprise itself according to a predetermined formula, without therefore involving the government, but not in all cases. Small enterprises without full-time accountants might lack the necessary expertise and thus require the services of private, or governmental, value assessors. And, in any case, a certain amount of governmental auditing will be necessary for keeping mistakes and deliberate miscalculations within reasonable bounds. And some means will have to be established for interested parties to appeal any determinations with which they disagree. I do not know how costly and time-consuming all of this will be, but I suspect that, overall, it will amount to a red-tape problem of significant proportions, yet not, perhaps, as significant as the red-tape problem of worker-control socialism.

Let us now take stock. In favour of worker-control socialism is that it avoids the special red-tape problem of worker-control capitalism, although it has its own, probably even more serious red-tape problem. Also, by means of its control over investments, the government with worker-control socialism does have a tool for fighting unemployment that is unavailable to governments with worker-control capitalism (Sect. 4.2). How significant this tool is, I do not know. Considering the democracy and incentive problems discussed above, this advantage may not be as significant as one might think. Next, it may be argued that worker-control socialism has an edge over its capitalist counterpart in the control of externalities (Sect. 4.2), but the dismal record of socialist countries so far in controlling pollution suggests that, if it does have an edge here, it is not a substantial one. Finally, worker-control socialism furthers greater equality of wealth by eliminating one major source of unequal wealth; namely, private investment. But this advantage is countered by the fact that worker-control capitalism furthers greater equality of wealth by spreading the wealth from investments out widely among all workers, rather than channelling it largely towards the wealthy, as does traditional capitalism.

In favour of worker-control capitalism, on the other hand, is that it avoids all six of the serious problems discussed above (Sect. 4.3) that are created by the socialized investment feature of worker-control socialism. These advantages are indeed not insignificant. Moreover, workers who themselves own the capital assets upon which they work can be expected to take better care of these assets than do workers who work with capital assets that are socially owned. Like it or not, it is only human nature to treat more carefully, repair more conscientiously and, in general, take better care of that which is one's own rather than society's. And, of course, the more successful an economy is in preserving its capital assets, the greater, ultimately, will be its productivity.

All in all, the advantages of worker-control capitalism are more numerous and clear-cut. Of the two varieties of worker control, it is, therefore, the more attractive alternative.

4.6 COMPARED WITH TRADITIONAL CAPITALISM

Let us turn, finally, to a comparison of the advantages and disadvantages of worker-control capitalism with those of traditional capitalism, the sort of capitalism that is found throughout much of the world today.

To begin with, worker-control capitalism will, for two reasons, yield a more equal distribution of income and wealth. As explained earlier (Sect. 4.2), managers that have to face workers in periodic elections will be unlikely to appropriate for themselves quite the large sums that those of traditional enterprises do. Also serving as an equalizer will be the fact that investment returns will be spread out among numerous workers, rather than falling largely in the hands of a relatively few rich. And with a more equal distribution of income and wealth comes a more equal access to basic necessities.

All of this, in turn, will mean more equality of opportunity as well. And contributing to equality of opportunity in general will be the greater opportunity most will have under worker control to express opinions effectively and exercise some influence at their workplace.

Next, as explained earlier (Sect. 4.2), the workplace democracy of worker control will tend to generate managers who cater more than do those in traditional enterprises to the well-being of workers. The potential benefits of this for workers are many; they include better, safer working conditions, jobs that are less tedious, more challenging, and a more satisfactory balance for workers between work and leisure. Take, for example, a more satisfactory balance between work and leisure. Since, with worker control, management is answerable to the workers through periodic elections, management will have an incentive they probably would not have with traditional capitalism to experiment with such worker-oriented arrangements as allowing a husband and wife team to split, fifty-fifty, a single job between themselves, allowing people to work (and be paid for) only every other day if they so chose, allowing them to work only 6 hours a day, allowing the option of longer vacations at reduced pay, and perhaps even allowing workers to have 'sabbaticals' at full, or half pay, so as to give them the opportunity to develop a new skill or pursue some creative project. (With, however, a job split fifty-fifty, or with a worker working only every other day, some sort of accommodation in the amount the worker is to be paid off upon leaving has to be made. Perhaps in these cases the worker can retain full voting rights, but have only 50 per cent 'pay-off' rights to match his or her 50 per cent working time.)

Moreover, the potential gains for workers from workplace democracy go beyond such concrete benefits as safer, more pleasant working conditions, somewhat less tedious, more challenging work, and a more favourable balance between work and leisure. These potential gains include also the more abstract benefits of more self-determination or autonomy, which, in itself, is likely to result in more satisfaction and fulfillment from one's work.

Invariably I am told by those who defend traditional capitalism that, although these many benefits for workers are a good thing, they can be achieved just as well within traditional capitalism (supposedly through such means as employee stock option plans, a token worker on the board of directors, autonomous work groups, and the like). I wish this were true, but I doubt it. Certainly improvements for workers within traditional capitalism are possible, but not likely to the extent that they are with worker control. It is simply a matter of institutional structure. Traditional corporations are structured so as to motivate managers to benefit, first and foremost, the owners of the corporation, the shareholders. So their concern for the well-being of workers is, in effect, limited largely to the extent to which worker and owner interests happen to coincide. It is not that managers are necessarily evil people who do not care about the interests of workers. It is merely that, according to the institutional structure of traditional capitalism, it is the owners, not workers, to whom they are accountable. So it is owner, not worker interests to which they are likely to give priority; if they do not, they will soon find themselves replaced by managers who do. And in the case in which the managers themselves happen to be the owners, they are accountable to no one but themselves, which again means that they have little motivation to give priority to worker interests. Worker control, on the other hand, by making managers accountable to workers, is specifically structured so as to motivate managers to give worker interests priority.

Consider, once again, the analogy between democracy in the workplace and political democracy. A totalitarian government may be, in principle, just as capable as a democratic government of acting in the general welfare. But we know that, nevertheless, a totalitarian government is less likely than a democratic one to actually do so. Why? It is, again, a matter of institutional structure; a political democracy is structured so as to make the government accountable to the people as a whole through democratic elections. Therefore, if a democratic government strays too far from the general welfare, it is likely to find itself out of power. To be sure, all democracies are plagued, to a greater or lessor extent, by disreputable politicians who cater to special interests, but if the democracy is reasonably healthy, the government will nevertheless cater to the general welfare for the most part (see Sect. 3.6). Likewise with the managers of a business enterprise; they will, for the most part, cater to the interests of those to whom, structurally, they are made accountable. According to how worker control is structured, managers are made accountable to the workers; therefore it is the interests of the workers to which they will, for the most part, cater. This is why traditional capitalism is unlikely ever to match worker control in being beneficial to workers.

At this point in the discussion, the defender of traditional capitalism may tell us that the difference in structure between the systems makes no real difference since, 'in the long run', worker and owner interests always coincide. Their interests always coincide, so the argument goes, because, if managers cater to the interests of workers, then workers will be happy and productive, and if workers are happy and productive, then owner interests (which take the form of owner profits) will be maximized. So, it will be concluded, the potential of traditional capitalism for benefiting workers is just as great as that of worker control.

The most serious flaw in the claim is, of course, that the alleged identity of interests upon which it is premissed is most definitely false. Only the most foolish of idealists could possibly believe that, with traditional capitalism, worker and owner interests always, or even nearly always, coincide. The only way for there ever to be an identity between worker and owner interests is for workers and owners to be, literally, one and the same people. And the only way for workers and owners to be one and the same people is, of course, to adopt worker control.

A significant reduction in how unethically and insensitively workers are often treated these days is itself sufficient justification for worker control, provided there are no costs that outweigh this benefit. But unethical and insensitive business practices extend not just to workers, but to consumers and the general public as well. Take, for example, the deliberate dumping of deadly chemical waste throughout the United States, the crass refusal to redesign the Ford Pinto so as to save many lives, and the recent, unscrupulous practices in the savings and loan industry. Can worker control bring about greater sensitivity to the needs not only of workers, but of consumers and the general public as well?

Perhaps. Take the ethically insensitive practices noted above. Are people in business really as insensitive to ethical considerations as these examples suggest? Probably not. As James A. Waters argues, most people in business are decent enough, but certain 'organizational blocks' preclude or discourage them from doing anything about the many unethical practices within their organizations (Waters 1989). These organizational blocks include strict lines of command, task-group cohesiveness, ambiguity about priorities, separation of decision-making, and division of work. What is needed more than anything else to counter these organizational blocks, Waters concludes, is an organizational atmosphere in which 'questioning, confrontation, and controversy' are valued, an atmosphere in which open discussion of, and challenges to, current practices are encouraged (1989: 159). Nothing might be more successful in creating just such an atmosphere than workplace democracy. What, after all, might be more conducive to an atmosphere of questioning, confrontation, and open discussion than for management to have to justify their practices periodically in a fair and open election in which they are opposed by candidates eager to uncover any shortcomings at all within the organization, be they ones pertaining to business or morality? (Mere shareholder 'voting', as currently practised, does not, for obvious reasons, create this atmosphere.) Consider, once again, the analogy with political democracy. On the whole, democratic governments tend to behave more ethically not only towards their own constituencies than do totalitarian ones, but also towards other countries. Although the treatment of other countries by democratic governments, such as those of the United States and Great Britain, hardly qualifies as saintly, their treatment nevertheless compares favourably to the excesses of many totalitarian governments, such as Hitler's Germany and Stalin's Russia. The pursuit of unethical practices is simply more difficult in a democracy than in a totalitarian system, as was demonstrated by the Vietnam War. This is because of the constraints under which democratic, but not totalitarian, governments must operate. These include not only free elections and constitutional checks and balances, but also free and uninhibited discussion and questioning of all major policies by the press and general public. It is not unreasonable to expect that, likewise, the constraints of democracy in the workplace will make unethical practices more difficult for businesses; that, because of the constraints of open discussion, businesses will tend to behave more ethically not only towards their own workers, but towards consumers and the general public as well.

On the other hand, much of these gains from worker control—gains in equalizing wealth and access to necessities, in equalizing opportunities, and in freedom, autonomy, and morality—will be offset if this system proves to be far less productive than traditional capitalism. Equal access to necessities means little if everyone is equally lacking in necessities. Opportunities and freedom suffer as well with a serious decline in productivity. But is a serious decline in productivity with worker control a likely prospect? There are indeed certain reasons for suspecting so. Let me now present these reasons, beginning more or less with the ones that are least important and progressing towards ones that are more important.

Some claim that, with the democratic format of worker-controlled enterprises, efficiency will be lost through workers spending too much time in meetings, bickering about this matter and that, and taking innumerable votes. This, however, will only be a serious problem with larger enterprises that attempt to operate as a direct democracy. Most larger enterprises, so as to avoid this very problem, will no doubt operate as a representative democracy instead, in which case a loss of efficiency through numerous meetings, votes, and so on, should be no serious problem.

Others claim that the democratic format of worker control will lead to frequent changes in top management, and thus frequent changes in policies and long-run plans, which will be inefficient. The idea is that, with frequent changes in policies and long-run plans, an enterprise may not be able to go far enough in any one direction to accomplish much.

But whether changes in top management, and, thus, in policies and long-run plans, really will tend to be unduly frequent with worker control is questionable. And, in any case, any such loss in efficiency must be weighed against the gain from being able to pry inefficient management from their positions more easily.

Perhaps a more credible reason for a loss in productivity with worker-control capitalism is the red-tape problem mentioned earlier. This is the problem of having to assess, periodically, the value of all enterprises, and of creating a means for appealing these assessments.

Still another problem with worker control, one that applies equally to either its socialist or capitalist version, is that of 'political' discrimination within the workplace. If management must face workers periodically in workplace elections, but management also has the power to decide upon hirings, firings, salaries and promotions, then management can always use this power to increase their chances in the next such election. Even though elections are to be by secret ballot, prior to the elections workers will likely be talking company politics with one another and aligning themselves with one faction or the other, all of which should make it fairly easy for management to determine who the leaders of the opposition are. Such workers can then be fired, with workers more friendly to management being hired to take their place. Or management can entice people to support them by rewarding their most ardent supporters with promotions and salary raises, and threatening to demote those who do not support them. In any worker-controlled enterprise, management's temptation to manipulate hirings, firings, promotions, and raises for their own political advantage will be great, and to the extent that such political discrimination does occur, efficiency, and thus productivity, will suffer. To be sure, traditional enterprises suffer from certain kinds of political manoeuvrings of their own, but not from exactly this kind.

At least a partial remedy for political discrimination within the workplace is simply to make it unlawful, just as racial and sexual discrimination within the workplace is unlawful in the United States today. But then this is where red tape and bureaucracy enter the picture again. For a law prohibiting political discrimination within the workplace to be effective, any worker must have ready means of appealing any hiring, firing, promotion, or salary decision he or she thinks is politically motivated. But given that many workers are likely to rationalize any such decisions that go against them as having been politically motivated, and that others are likely to appeal just to try to get even, or merely on the off chance of winning, such appeals might become numerous. And any bureaucracy large enough for handling so many appeals might end up being another costly drain upon the economy. Moreover, if such appeals were numerous and successful enough, management might then become the ones that were intimidated; they might become unwilling to let incompetent workers go and to make other such hard, yet necessary decisions, for fear of being accused of political discrimination. This, again, would hurt efficiency.

On the other hand, these detrimental things might not happen after all— or, at least, not to an extent great enough to seriously affect productivity. We can only know after much more observation than so far of worker control in practice.

Another potential source of inefficiency and thus diminished productivity with worker-control capitalism (or socialism) is an alleged tendency of worker-controlled enterprises to react perversely to increases and decreases in the demand, and thus price, of their product (Ward 1967; see also Meade 1980: 94–5). It can be shown mathematically that, at least in the short run, as the demand and price for a worker-controlled enterprise's product increase, and thus so does income per worker, a still greater (although not much greater) increase in income per worker could be achieved by laying off some workers. Conversely, as demand, price, and thus income per worker decrease, this decrease in income could be minimized slightly by hiring some new workers. The general idea is this. First, let us take the 'ideal' number of workers in an enterprise to be the number at which income per worker is maximized. Next, let us assume that the productivity of each additional worker in an enterprise will be slightly less than that of the previous worker—that is, let us assume declining marginal productivity at a more or less constant rate. (This assumption is generally true for the short run during which the enterprise's equipment is fixed.) If declining marginal productivity were the only factor determining what the ideal number of workers were, then obviously the ideal number would be one. In other words, declining marginal productivity exerts pressure in favour of there being fewer workers. Finally, let us assume the enterprise has a certain amount of fixed expenses, ones such as repayments of loans, the maintenance of depreciation funds, and so on, expenses that do not vary with the number of workers there are. If fixed expenses were the only factor determining the ideal number of workers, then the ideal number would be infinity, since, obviously, the more workers there are over which any fixed expenses are to be spread, the higher the average income per worker will be. In other words, fixed expenses exert pressure in favour of there being more workers. Now notice the following. As demand, price, and thus income per worker *increase*, the importance of fixed expenses for determining the ideal number of workers *decreases* relative to the importance of declining marginal productivity (the higher the price paid for the enterprises' products, the more the loss in income that any declining marginal productivity entails). And as the relative importance of fixed expenses thus decreases, then so does the ideal number of workers. If, instead, demand, price, and thus income per worker *decreases*, then the relative importance of fixed expenses, and thus the ideal number of workers, *increases*. So, theoretically, there will be a tendency to lay off workers in times of increasing demand, and to hire them in times of decreasing demand. One of the main benefits of a free-market economy, however, is supposed to be how well it responds to people's wants and desires. But if, whenever people want more of a product, as shown by increased demand, the economy, due to these lay-offs, responds by producing less of it instead (and vice versa whenever people want less of a product), then the economy will hardly be responding well to people's wants and desires; it will be responding perversely.

One might wonder, however, whether this so-called perverse response is really so bad. If, whenever there is a decrease in demand resulting from a recession, worker-controlled enterprises respond by hiring more workers, then maybe this will serve to keep the economy more stable, to minimize the destructive consequences of business cycles.

But the desirability of this so-called perverse response is, I think, a moot point, for I doubt if worker-controlled enterprises really will, typically, respond this way. With an increase in the demand for their products, workers' incomes will have gone up independently of any lay-offs. But after their incomes have thus gone up, I doubt if these workers will then be so greedy as to lay off some fellow-workers merely to increase their incomes just a little more. Surely the solidarity among workers in a worker-controlled enterprise will typically preclude any such betrayal as this (see Vanek 1977: 20; Schweickart 1980: 72). Of course *hiring* workers in response to a *decrease* in demand involves no betrayal of fellow-workers. Yet the amount gained by thus hiring workers in response to a decrease in demand would probably be so small, relatively speaking, that most enterprises would not even bother—although, in the case of hiring workers, a tendency to do so would not, as suggested above, be such a bad thing anyway.

But, although worker-controlled enterprises may not *lay off* workers in response to increased demand, they, contrary to traditional enterprises, are not likely to *hire* workers either. They are not likely to hire workers in the short run, for largely the reasons just considered. In the long run, worker-controlled enterprises will have reason to hire new workers and expand if, by doing so, they can achieve economies of scale. But once they have exhausted their potential for economies of scale, they no longer will be likely to hire new workers and expand even in the long run, while traditional enterprises may still do so. The reasons that, having exhausted its potential for economies of scale, a worker-controlled enterprise is less likely to expand, in the long run, than a traditional one is simple enough. A traditional enterprise, even without any economies of scale, can, by expanding, often increase its owner's profits, but without any economies of scale a worker-controlled enterprise can never, by expanding, increase income per worker. Schweickart illustrates this with the example of a hamburger stand employing 20 people (1980: 72). Say this enterprise is a traditional one, and that it makes $20,000/year profit for its owner (i.e. after salaries and other expenses have been paid, the owner has $20,000 left over). If, by expanding her enterprise through hiring another 20 people and opening an identical stand on the other side of town, the owner could make another $20,000/year in profit, she would have every reason to do so. But say that, instead, the stand is a worker-controlled enterprise. The $20,000 in profits (the amount by which their overall revenue exceeds their 'base' salaries and expenses) will then be divided among the 20 workers. But if, through opening an identical stand on the other side of town, they could make another $20,000 in profits, they still would not have increased their income per worker, for they would then have another 20 workers among whom this $20,000 would have to be divided. Thus, contrary to the owner of the traditional firm, they would have no financial incentive for expanding. Given that the same will be true of any worker-controlled enterprise under conditions of constant returns to scale, it is questionable whether an economy consisting primarily of worker-controlled enterprises will therefore have as much new investment and growth as an economy consisting primarily of traditional enterprises.

Schweickart suggests that this lack of incentive to expand can be compensated for in two ways. One way is merely by explaining to workers the importance of expansion to the economy as a whole, and then relying upon the workers to expand their enterprises solely for the good of the economy (1980: 73).

This way is basically worthless. Mere 'jawboning', such as this, has not worked well in the past in other contexts, and cannot be relied upon to work well in the future. For one thing, jawboning appeals give rise to serious 'assurance' problems (Sect. 2.5).

The other way Schweickart suggests that a lack of expansion can be compensated for is by new ventures. For example, rather than the old hamburger stand doubling in size as in a traditional economy, a new, entirely independent stand might start up instead. This second potential way of compensating is far more promising. Indeed, the version of worker-control capitalism outlined here is probably better suited for compensating by means of new ventures than is Schweickart's own version of worker control. In the first place, this version of worker control allows new ventures in which an entrepreneur is permitted to keep total control of the venture himself. This version of worker control, in other words, leaves room for traditional entrepreneurs. So even though the old hamburger stand in the example above may have no incentive to expand, a traditional entrepreneur will have a $20,000/year financial incentive for starting a new, competing stand. Moreover, with this version of worker control, new ventures are encouraged as well by favourable credit terms for entre-preneurs willing to adopt a worker-control format. Thus, if the opportunity to open this new hamburger stand were not snatched up by a traditional entrepreneur, then others—perhaps recent graduates in need of employ-ment—might take advantage of these credit terms and open it as a new worker-controlled venture. In general, the version of worker-control capitalism outlined here is not subject, as is Schweickart's socialist version, to the incentive, democracy, and (socialist) red-tape problems considered above. So investments in new ventures may well take up any investment slack resulting from a lack of incentives for expansion—just as Schweickart suggests. And the version of worker control outlined here my well take up this slack even better than would the version Schweickart himself advocates.

And, incidentally, special measures for encouraging investments in new ventures are available. Prominent among these is, as Schweickart suggests, making the profit-per-worker ratio of all worker-controlled enterprises a matter of public record, so that the most promising areas for new ventures can thereby become well known (1980: 73).

On the other hand, it is generally more difficult and time-consuming for new enterprises to start up than for existing ones to expand. So relying upon the creation of new enterprises could prove to be a more cumbersome, less flexible, and generally less effective way of bringing about economic growth than merely through expanding existing enterprises.

Moreover, a worker-control economy in which more investment in new enterprises makes up for less investment in the expansion of old enterprises will be characterized by a greater number of independent, but smaller enterprises than in traditional capitalism. A greater number of independent, but smaller enterprises is not, of course, necessarily bad. It may bring about a more competitive economy, along with the increases in efficiency and innovation that come with more competition. Competition in the hamburger business, and thus perhaps efficiency and innovation, should be greater in the town with two entirely independent hamburger stands each employing 20 people than in the town with, instead, two branches of the same hamburger enterprise which employs 40 people overall. And the smaller the enterprise, the more intimacy there is likely to be between managers and (other) workers, and the more a worker is likely to be able to participate actively in the decision-making process, and feel like he or she is a significant part of the enterprise, rather than just a 'small cog in a big machine'.

But there may be one major drawback to an economy that tends towards many, smaller enterprises rather than fewer, larger ones. Serious research and development is generally quite expensive; therefore, it is argued (e.g. Galbraith 1967), only the largest enterprises have the financial resources necessary for it. Would then the smaller enterprises of worker control, with their more limited financial resources, mean less research and development? I do not know, but if it did mean less research and development, then, since research and development is often the impetus for investment and growth, this would mean less investment and growth.

Finally, research and development, and much new investment, entails sacrificing current profits for the sake of greater profits in the future. Those who control traditional enterprises, the major shareholders of corporations and the owners of other large enterprises, are, typically, wealthy people; they can afford to sacrifice current profits for the sake of even greater profits in the future and, therefore, often encourage research, development, and new investment. But would ordinary workers, who control worker-controlled enterprises, be equally willing to sacrifice current profits—which would mean a cut in their relatively modest, current incomes—for the sake of greater profits in the future? If the unwillingness of voters in Western political democracies to favour higher taxes for the sake of the future is any indication, then the answer is no, workers will not generally be as willing as traditional owners to sacrifice current profits for the sake of greater future profits. After all, workers are not normally wealthy people, and therefore a cut in current profits would hurt them far more than it does the wealthy owners of traditional enterprises. So even if worker-controlled enterprises—or enough of them—did have the financial resources necessary for serious research, development, and new investment, it might be less likely than with traditional enterprises that they would choose to use these resources in this way. To put it in economists' terminology, the 'time horizons' of worker-controlled enterprises might tend to be shorter than those of traditional ones. (Do not, however, confuse this time-horizon problem with the even more serious one referred to in Sect. 4.4 that this version of worker control avoids.)

In reply to these doubts about the time horizons of worker-controlled enterprises and their potential for research and development, it might be argued that, if research and development did prove to be insufficient in a worker-control economy, then the government could always take up the slack through wide-ranging subsidies of research and development. But wide-ranging governmental subsidies are not an altogether satisfactory substitute for too little research and development emanating from private enterprises; subsidies raise troubling questions. Will government subsidize the research and development that is most compatible with people's market preferences, or will it follow an agenda of its own (one, say, heavily oriented towards the defence industry)? Will governmental subsidies adequately represent interests that are not among the most popular? Will not research and development controlled largely by politicians tend to cater too heavily to special interests? Will short-sighted taxpayers support research and development to a sufficient degree? And will the bureaucracy necessary for administering this research and development be too costly and cumbersome? In short: if an economy relies too heavily on the government for research and development, this raises the very same problems of élitism, tyranny of the majority, democracy, and red tape that we encountered earlier in connection with socialized investment in general.

So far, then, we have found reason to question the adequacy of the time horizons, and new investment potential, of worker-controlled enterprises. If, indeed, worker control will not be as successful in generating new investment, then this is still another reason, probably the most significant reason by far, why, in the long run, worker control may not be as productive as traditional capitalism. But, before coming to any overall conclusions, we must also see what, with respect to productivity, may be said in favour of worker control.

Let us begin with the question of time horizons. As we saw, it would appear that workers, with less wealth on the average to 'play' with than traditional shareholders, would therefore be less willing to trade immediate returns for investment in the future. But now consider the other side of the argument. Even with worker control, the incomes of ordinary workers are likely to be substantially less than those of management. Yet everyone's future share in the ownership of the enterprise, from the lowest of ordinary workers to the highest of managers, is exactly the same. What this means is that if any extra revenue the enterprise has earned is distributed as income, ordinary workers are likely to get substantially less per capita than management; yet if this revenue is instead ploughed back into the business thereby increasing the enterprise's value, the ordinary workers' per capita share in this increase and management's will be exactly the same. Thus it may well be much more in the ordinary workers' financial interests for any extra revenue to be plowed back into the business, that is, invested. This financial advantage may tend to counteract somewhat the natural reluctance of those with only modest incomes to trade immediate returns for investments.

Moreover, traditional capitalism, at least as practised in the United States today, is faced with a time-horizon problem of its own. The time horizons of traditional enterprises today tend to be too short for a number of reasons. First, there is the constant threat of hostile takeovers. This threat puts considerable pressure upon managers to come up with results immediately, or else. Sinking funds into long-term research and development at the expense of current dividends is liable to result in a temporary devaluation of the company's stock, thus making it a prime target for hostile takeover (Thurow 1985: 150). Moreover, those enterprises involved in successful takeovers often do not sink substantial funds into long-term research and development either, in part due to the necessity of repaying the huge loans that were used to finance the takeover (Reich 1989: 16–17). Other enterprises are borrowing heavily, and cutting back on long-term research and development, so as to retire equity in the hope of fending off takeovers. In short, a business climate featuring hostile takeovers and tricky share manipulations of all sorts is hardly one in which long-term research and development is likely to thrive.

With worker-controlled enterprises, on the other hand, only current employees are allowed to own a voting share of the enterprise. Thus, obviously, worker-controlled enterprises are free from the manipulations of share ownership that make hostile takeovers possible. Accordingly, they are also free from the various takeover-related pressures that are among the reasons for the inadequate time horizons of many traditional enterprises.

A second reason for the short time horizons of traditional enterprises these days is the pressure for immediate results exerted by shareholders. As Charles Reich points out (1989: 17), 70 per cent of corporate stock is bought and sold these days by professional portfolio managers of mutual funds, pension funds, and insurance companies. These professional portfolio managers, in order to attract and keep clients and thus their own jobs, are under enormous pressure for immediate results, and much of this pressure is, in turn, transferred to corporate managers. But worker-controlled enterprises should not, for the most part, be subject to these particular pressures, since they are not financed by the voting stock in which these portfolio managers, for the most part, deal.

A third reason for the short time horizons of traditional enterprises today is that, as a result, among other things, of hostile takeovers and institutional shareholder pressure to produce immediately or else, the average term of office for today's chief executive officers is only about four years (Reich 1989: 17; see also Thurow 1985: 150). Realizing that he is likely to be in office only a few years, what CEO, Thurow asks, is therefore going to take a long-term view? 'Only a saint,' Thurow answers, 'and there aren't many saints among top management' (1985: 150). The upshot of all of this, Charles Reich concludes, is that among the three major components of today's traditional corporate enterprises—shareholders, management, and (lower-level) employees—it is actually the employees that, typically, have the most interest in the *long-term* well-being of the corporation. 'Employees', he writes, 'must live with the consequences of declining long-term productivity within an industry and a region; investors and managers often can bail out long before' (1989: 17). But if Reich is correct, then perhaps is it not as much enterprises controlled by workers that have a time-horizon problem as it is traditional enterprises, subject as they are to the pressures of hostile takeovers, and controlled by portfolio professionals and short-term managers.

Let us turn next to another possible advantage of worker-control capitalism with respect to productivity. All the workers in a worker-controlled enterprise own equal shares and are thus equal partners in a joint endeavour; as such, they are likely to feel a certain solidarity with one another that does not exist in traditional enterprises. Partners normally do not lay off one another merely because of unfavourable, but temporary circumstances (a recession) for which no one is responsible. Rather partners generally stick together and share the burden until circumstances become favourable again. Therefore, during recessions, the workers in worker-controlled enterprises, by agreeing to less pay across the board, will likely share the burden among themselves instead of placing the entire burden upon a minority of workers by laying them off, as is the practice in traditional enterprises. So, during recessions, a worker-control economy is likely to suffer less from unemployment and the vicious downward spiral of the economy to which unemployment contributes.

A possible down side of this is that worker-controlled enterprises may tend to keep workers beyond the point at which, for the sake of efficiency, they should be kept (Elster 1989: 94, 106). But, as both common sense and the current corporate practice in Japan of 'lifetime employment' suggest, a tendency to keep workers too long is likely to be far more in the interests of both workers and a healthy economy than is the tendency in most current capitalist economies to lay them off too quickly. And, after all, a free-enterprise economy has a tendency built into it that counters any tendency to keep workers too long: namely, the tendency for incomes in an enterprise with too many workers to go down relative to incomes elsewhere, thereby providing the workers with an incentive to leave on their own accord.

Still another possible productivity advantage of a worker-control economy is that (except in the relatively few enterprises not yet bought out) there will no longer be a need for labour unions. The purpose of labour unions is to represent the workers in negotiating with owners, but in a worker-controlled enterprise, the workers themselves are the owners, and thus a union will be unnecessary. One can certainly sympathize with the goal of labour unions—namely, that of alleviating the triple burden from which workers suffer under traditional capitalism; indeed, this is the very goal that motivates worker control (see Sect. 4.1 above). But the stark fact is that labour unions are often bad for the economy. They hurt productivity with prolonged strikes. Moreover, to the extent that they do succeed in pushing wages up, they do so only at the cost of greater unemployment (as employers no longer can afford as many workers) or at the cost of greater inflation (as employers raise the prices of their products in order to cover the costs of these higher wages), all of which also hurts productivity. And, ironically, this unemployment and inflation are most contrary to the interests of the very group—the workers—whose interests unions are dedicated to promoting. Worker control, since it puts the workers themselves in charge, promotes their interests more directly, and thus effectively, and does so without the counter-productive strikes, unemployment, and inflation generated by unions. And, to the extent that a worker-control economy thus avoids these counter-productive consequences of unions, it will, of course, be more productive.

We come finally to what may be the least tangible, but most important productivity advantage of worker-control capitalism. Since, with worker control, management typically caters more to the interests of workers than with traditional capitalism, workers, typically, will be more satisfied. And, typically, workers will also be more motivated to work hard and be creative, since each shares equally in the success of the enterprise (salaries are not equal, but shares in the ownership are). The more satisfied, the harder working, and the more creative workers are, then, other things being equal, the more productive they will be. So if, throughout the entire economy, workers are generally more satisfied, harder working, and more creative, a far from insignificant gain in productivity may well be the result. (On the importance, in particular, of there being more creativity throughout the entire workplace rather than its being concentrated largely among traditional 'entrepreneurial heroes', see Reich 1989: ch. 9.) As Lester Thurow writes: 'America's biggest handicap is found in its inability to generate an environment where the labour force takes a direct interest in raising productivity' (1985: 148–9). Nothing, I suggest, has greater potential for generating such an environment than worker control.

A contrary view is defended by Armen Alchian and Harold Demsetz. They argue that maximum input from workers can only be achieved within a traditional capitalistic format. The general idea seems to be this. Many productive activities are performed most efficiently by team effort rather than individually. But compensation tied to *team* performance is ineffective; effective incentives to work depend upon there being a close correspondence between *individual* performance and compensation. Since a worker-controlled enterprise constitutes a 'team' in the Alchian and Demsetz's sense and, typically, each worker's income in worker-controlled enterprises is tied to overall enterprise performance (at least in part), it follows that Alchian and Demsetz's argument applies to worker control. With worker control, any decrease in enterprise income resulting from a decrease in a worker's performance is, largely, spread among all the workers, rather than being borne entirely by that one worker himself. Likewise, any increase in enterprise income resulting from an increase in a worker's performance is, largely, spread among all the workers. In other words, each worker can decrease or increase his performance without a proportionate decrease or increase in his own income. But if so, each worker has a motive for shirking—that is, not doing his best—thereby becoming a 'free rider'. By shirking, a worker can gain more by way of 'leisure' than he loses by way of the relatively small change in income that (because income is tied to team performance) he personally will suffer. The only way to keep shirking within reasonable limits, Alchian and Demsetz say, is by effectively 'monitoring' team members—that is, by carefully policing them, and firing (or demoting) those who shirk.

So, the advocate of worker control might reply, why not simply assign the job of monitor to one or more of the workers? The problem, Alchian and Demsetz claim, is that, if compensation is still tied to team performance, any increase in income from the monitor's performance will, once again, be spread among the team as a whole, thus providing even the monitor with a motive for shirking. The only solution to this problem, Alchian and Demsetz conclude, is to provide the monitor with sufficient job incentive by allowing him to keep, personally, the gains in income resulting from his monitoring rather than spreading the gains around. In other words, the monitor must be able to keep the team's 'residual' income. But the only way for the monitor to be able to keep the team's residual income is for him to be the owner of the enterprise rather than just another worker. And the only way for him to be owner is for the enterprise to be a traditional one. So, Alchian and Demsetz conclude, only with a traditional enterprise is the shirking, or free-rider, problem solved. Therefore, they say, traditional enterprises are more efficient, other things being equal, than worker-controlled ones.

There are many problems with this argument. First, there is no reason, in principle, why a monitor in a worker-controlled enterprise (as it is understood here) could not have both the powers to fire and demote, as well as a claim to the residual income even—if indeed a claim to the residual income was necessary in order to entice effective monitoring. But, secondly, that a claim to the residual income is necessary to entice effective monitoring is implausible. A monitor (i.e. foreman) can be sufficiently motivated by being paid more than those he is monitoring, provided he also is under the threat of being demoted if he does not perform adequately and is himself being effectively monitored.

But, it may be asked, how can the monitor (i.e. foreman) be effectively monitored within a worker-control format? The answer is obvious: in exactly the same way that he is monitored in a traditional enterprise, by management. But, in a worker-controlled enterprise, how then can management be effectively monitored? The answer, again, is obvious: by the workers, to whom management is answerable through workplace democracy. In short: with worker control, there is not a *hierarchy* of authority in which those at the top, being unmonitored, must therefore be motivated by residual income. Instead, there is a *circle* of authority in which no one goes unmonitored and, with no one going unmonitored, there is thus no need for motivation by means of residual income.

Can workers, however, really be effective monitors of management, as effective as the shareholders (i.e. 'residual claimants') of a traditional enterprise? They cannot, it may be replied, since workers do not have either the knowledge or the incentive to be as effective as shareholders (cf. Jensen and Meckling 1979). They do not have the knowledge because they are not, like shareholders, guided by sophisticated, professional financial analysts. And they do not have the incentive because, once again, they do not, like shareholders, get the residual income.

But to say that workers have less incentive is, frankly, absurd. First of all, workers in a worker-controlled enterprise are, of course, just as much claimants of the residual income as are shareholders in a traditional enterprise. And, in addition, a worker's share in a worker-controlled enterprise is, typically, far more important to him than is a shareholder's share in a traditional enterprise. A shareholder's share is likely to be just a small part of his overall, diversified portfolio and, in any case, the success of the enterprise will usually have little or nothing to do with the shareholder's lifework. A worker's share in a worker-controlled enterprise, on the other hand, is likely to be the worker's main, or even only, investment in life and the success of the enterprise will probably have a huge bearing upon the worker's lifework. Thus the incentives for workers in a worker-controlled enterprise to monitor management successfully will hardly be less than that of shareholders in a traditional enterprise. On the contrary, the workers' incentives typically will be far greater.

Consider now the workers' knowledge. Workers, typically, are in much closer proximity to the affairs of the enterprise and are more personally acquainted with the strengths and weaknesses of the managers than most professional financial analysts ever could be. This greater, first-hand knowledge is surely one point in favour of workers. Admittedly, however, the professionals have greater training and skills in evaluating the financial performance and prospects of an enterprise than do workers. Managers in traditional enterprises will be influenced—i.e. monitored—by these professionals mainly to the extent that they need to gain access to capital finances. But, as Louis Putterman has pointed out (1984), the managers of worker-controlled enterprises (of the sort in question here) will finance capital in much the same way as traditional enterprises; that is, by means of bank loans, the sale of bonds, and maybe even the sale of non-voting equity in the enterprise, all of which will involve the approval of these same professionals. This, in turn, means that the managers of worker-controlled enterprises will be monitored just as knowledgeably by professionals as are the managers of traditional enterprises. Overall, therefore, the managers of worker-controlled enterprises will be monitored at least as effectively, and (since workers have a greater incentive to monitor than shareholders) perhaps more effectively.

So far, we have seen that Alchian and Demsetz fail to establish that worker-controlled enterprises cannot motivate workers as effectively as traditional ones. But it is not just that worker-controlled enterprises are equal in motivating capacity to traditional ones. They may be superior. Worker-controlled enterprises may motivate performance in a way that can never be fully duplicated by traditional enterprises: with worker control, the workers are not just hired hands for the benefit of others; the enterprise is *theirs*; if it prospers, it is *they* who prosper; if it fails, it is *they* who fail. This sort of identification of workers with their enterprise is something a traditional enterprise can never quite match, and it may have at least three important motivational consequences.

First, workers who identify with their enterprise will, more so than traditional workers, be motivated to monitor and preserve the enterprise's assets. The owners of traditional enterprises will be just as motivated to monitor and preserve enterprise assets, but owner-workers who are on-site, as with worker control, are in a better position to do so than traditional owners who (for the most part) are absent from the workplace. So, with respect to monitoring and thus preserving capital assets, it would appear that worker-control capitalism has an advantage not just over worker-control socialism (see Sect. 4.5) but traditional capitalism as well. (Incidentally, it has also been argued that, because they must actually live in the environment surrounding their enterprise, on-site worker-owners will therefore tend to be more concerned than are the distant, absentee owners of traditional capitalism with the preservation of this environment.)

Secondly, the identification of workers with their enterprise provides them with an incentive not just to 'get by', but to go beyond ordinary duty; to be creative, take the initiative, think, make suggestions, and expend greater effort. These inputs cannot be motivated merely by punishment for non-performance; that is, merely by monitoring. These inputs can only be motivated by rewards for performance, and by the workers' identifying their well-being with that of their enterprise to an extent they are unlikely ever to do with traditional enterprises, they become implicitly rewarded for these inputs in a way that traditional enterprises are unlikely ever to be able to duplicate.

Thirdly, by workers in worker-controlled enterprises thus identifying their well-being with that of their enterprise, certain norms and values are, as a consequence, likely to prevail within worker-controlled enterprises that would not be likely in traditional enterprises. These, roughly speaking, are norms and values that encourage extra effort and discourage shirking and free riding. And (although economists appear not always to realize this) people are motivated by more than just money and leisure; they are also motivated by, among other things, norms and values. Otherwise, for even the most minimal law and order to prevail, there would have to be a policeman on every corner and in every house. Once the norms and values encouraging effort and discouraging shirking become 'internalized' by workers in worker-controlled enterprises, then the need for monitoring may well become less than it is in traditional enterprises. Moreover, what monitoring is needed may well, at least in part, be performed voluntarily by the workers themselves. Therefore the exact opposite from what Alchian and Demsetz argue may well be true; worker control may well be the format in which monitoring is most effective and efficient.

In sum: to disregard the potential gains in productivity from the motivational capacities of worker control would be a mistake. Those who support traditional capitalism over central-planning socialism do so largely because it spreads the incentive to expend extra effort and be creative around more widely than does central-planning socialism. Well, worker-control capitalism spreads this incentive around more widely yet.

What then should we conclude overall about the productivity of worker-control capitalism versus that of traditional capitalism? I do now know. As we have seen, with respect to productivity both systems appear to have their advantages and disadvantages. But nothing quite like the version of worker-control capitalism outlined here has ever been put into practice.

The Yugoslavian version of worker control, which had mixed results, perhaps came closest, but not close enough to draw any conclusions about worker-control capitalism from it. The Yugoslavian version of worker control has been described, and accurately I think, as 'an only half-reformed centralized socialist economy' (Putterman 1990: 177). For one thing, workers 'owned' capital assets collectively in the sense that they did not own individual shares that could be cashed in upon leaving the enterprise, nor were they allowed to liquidate these assets even after the claims of creditors had been satisfied. This gave rise to a serious 'time-horizon' problem (see Sect. 4.4). To try to solve this problem, the government, among other things, encouraged banks to supply large amounts of credit at excessively low or negative real interest rates, thereby causing high rates of inflation and preventing the rate of return from serving as an efficient allocator of scarce capital among competing projects. Moreover, the country's ethnic hostilities and political unrest placed additional burdens upon the economy, thereby making it difficult to sort out which problems were attributable to the economic system and which to these extraneous factors. (For more on the Yugoslavian system, see Lydall 1984.) Since, therefore, worker-control capitalism has yet to be tried, the productivity advantages and disadvantages of this system remain largely speculative.

4.7 A MIXED ECONOMY

In the light of everything, therefore, what should we conclude about the *overall* merits of worker-control capitalism relative to those of traditional capitalism? Let me put the question this way. Assuming, just for the sake of argument, that people in current capitalist countries were willing, should the legislation necessary for changing from traditional capitalism to worker-control capitalism be enacted at this time? My reply has to be no. The unanswered questions about the productivity of worker-control capitalism are too serious to warrant immediate, full-scale implementation.

Yet, on the other hand, the potential benefits from this system are great enough to warrant not rejecting it at this time either. Let me try to explain in somewhat different, and more general terms than I have so far what these potential benefits are.

I am not, quite frankly, impressed with much of Marxist social and political philosophy. Tied down, as Marxists are, by their 'mandatory' ideology and outmoded economics, their criticisms of current capitalism are often off target and their proposed cures worse than the disease, thereby perhaps causing many concerned people to be far more defensive, more protective of the status quo, than they otherwise would have been.

But one thing in Marxist philosophy interests me: the analysis of class conflict. One's class, in the Marxist sense of 'class', is determined not by one's wealth, but by how one is related to the means of production, to capital. Marxists tell us that capitalism is (roughly speaking) characterized by a class division between, on the one hand, those who own capital but, typically, do not work with it (the 'bourgeoisie' or 'capitalists') and, on the other hand, those who work with capital but, typically, do not own it (the 'proletariat' or 'workers') (Marx 1867). Capitalists are primarily interested in maximizing profits; workers are primarily interested in good wages and satisfying work. Thus the interests of capitalists and workers are very largely opposed; generally speaking, the more profits capitalists get, the less wages workers get, and the less safe, the less pleasant, and the less autonomous their work gets. Opposed class interests such as these, Marxists tell us, inevitably lead to antagonism, conflict, and exploitation of the less powerful class—in this case the workers.

The Marxist cure for all of this is to replace capitalism with a classless system, one, in other words, in which everyone is related to the means of production, or capital, in the same way. For traditional Marxists, this has meant replacing capitalism with central-planning socialism, for with central-planning socialism everyone is said to be related to capital both as owner in common with everyone else and as worker.

But it is precisely here where the Marxist argument went wrong. Far from being a classless system, central-planning socialism creates its own vicious class division: a division between those who *control* capital but do not work with it, and those who *work* with capital but do not control it. Those who control capital are, of course, the central planners and other government officials—we may call them the 'bureaucrats'—while those who work with it are, once again, the 'workers'. The gulf between bureaucrats and workers in central-planning socialism is enormous—even in cases where the bureaucrats are, in some sense, democratically elected. The bureaucrats are interested in maximizing their power, their control over the economy, while the workers, once again, are interested in good wages and satisfying work. And, as we have seen (Ch. 3), these class interests are no more compatible than are those of traditional capitalism. Maintaining control over the entire economy from the top is incompatible with the prosperity necessary for good wages and the worker autonomy necessary for satisfying work.

We come, finally, to what may be worker-control capitalism's most significant, potential advantage: the closest we can ever come to a classless society in the Marxist sense is through neither traditional capitalism nor central-planning socialism; it is through worker-control capitalism. For only with worker control capitalism are those who *work* with capital, those who *control* capital, and those who *own* capital one and the same: the workers. It is true that, with worker-control capitalism, those among the workers who become managers will have more control than other workers, but the gulf between them and other workers will not come even close to being as great as the gulf between workers and bureaucrats in central-planning socialism. So with worker-control capitalism, since, for the first time, work, control, and ownership will not be split among different groups, then maybe, for the first time, we will be largely free from class conflict and exploitation. This, along with the other potential benefits of worker-control capitalism, is what warrants giving this system a try.

But how, exactly, can we give it a try without invoking the unacceptable risks referred to earlier? Let us, I suggest, do the following. Full-scale worker-control capitalism, as outlined here, has three basic features: (a) legislative recognition of the worker-control format; (b) favourable credit terms for enterprises with this format; and (c) a buy-out requirement. Let us implement the first two of these three features, but not the third. Let us give the worker-control format legislative recognition similar to that presently given the 'corporation' format and make favourable credit terms available to enterprises with this format. This, especially if combined with the lifetime inheritance quota proposed in Chapter 6, should encourage far more worker-controlled enterprises than are found in traditional capitalist economies today. Yet, by not implementing the buy-out requirement, there will be little danger of moving too quickly to worker control and thus perhaps causing unacceptable harm.

The goal of implementing these two features is to bring about a 'mixed' economy, an economy in which a considerable number of worker-controlled enterprises compete head-to-head with traditional ones (cf. Krouse and McPherson 1986). With head-to-head competition between them, the answers to many questions about the relative merits of each should become clear. Not all the answers will become clear. Comparing alternative modes of economic organization is a difficult task fraught with methodological pitfalls (for a discussion of some of these pitfalls, see Elster and Moene 1989: 12–21). But if we proceed with caution, aware of the difficulties involved, we should eventually be able to answer enough questions so as to know at least whether to take yet another step towards worker control, or whether to retreat. If worker control lives up to its potential, the next step may then be to implement the buy-out requirement.

On the other hand, it may turn out to be best not to implement the buy-out requirement in any case. It may turn out to be best for a mixed economy featuring head-to-head competition between worker-controlled and traditional enterprises to continue indefinitely. One advantage of never implementing the buy-out requirement is that no enterprise will then ever by forced to take on a certain format and, other things being equal, the less coercion the better. Moreover, by not implementing this requirement, and thus never forcing any enterprises to become worker controlled, a much-debated problem is avoided: the problem of whether an exception to the requirement that all enterprises become worker controlled should be made for ones of a certain size, say very small enterprises, or ones of a certain type, say law firms and other professional associations, enterprises for which, arguably, a worker-control format is not suitable. Without the buy-out requirement, enterprises of a type that is not suitable simply will not, for the most part, become worker controlled, thereby avoiding the need for deciding from the start which types are suitable and which not. And, of course, if worker control does, in fact, live up to its potential, then most traditional enterprises may eventually become worker controlled even without the buy-out requirement.

But one final caveat. Widespread worker control, as beneficial as it may be, is nevertheless no panacea. Still other measures will be needed to adequately remedy some of the most serious deficiencies of traditional capitalism, which include extreme inequalities in wealth and opportunities. What these other measures should be is the topic of the final two chapters.

5
CAPITALISM WITHOUT POVERTY

5.1 INTRODUCTION

Capitalism today gives rise to greater wealth, and probably more personal freedom, than any other system the world has known. Realization of the ideals of freedom and productivity are capitalism's strengths.

Unfortunately, capitalism, with the vast inequalities to which it gives rise, does not do nearly as well in realizing two vital ideals of equality: equal access for everyone to the basic necessities of life and equal opportunity. Libertarians (arch-conservatives) and central-planning socialists (arch-liberals) often agree, curiously enough, about one basic premise. They agree that these two ideals of equality can never, through any governmental intervention, be realized much more fully than at present without abandoning capitalism.

What the libertarian and socialist disagree about is priorities. As between (a) the preservation of a viable capitalism and (b) a much greater realization of these two ideals of equality, the libertarian places higher priority upon the preservation of a viable capitalism. Governmental attempts to achieve a much greater access for everyone to basic necessities and far greater equality of opportunity, the libertarian tells us, only serve seriously to undermine freedom and productivity, and thus capitalism itself; let us therefore, the libertarian concludes, abandon any such attempts. For the socialist, on the other hand, these two ideals of equality have priority. Since, the socialist agrees, these two ideals cannot be realized to a much greater extent within a viable capitalist framework, let it therefore, the socialist concludes, be capitalism that we abandon; let us bring about a planned economy along with access for all to necessities and opportunities.

I tend to agree with the socialist's priorities; if the vast inequalities in access to necessities and opportunities that current capitalism breeds cannot be greatly alleviated within a capitalist framework then, I say, so much the worse for capitalism. But I reject the underlying premise with which both libertarian and socialist agree. Contrary to both libertarian and socialist, I believe that, through appropriate governmental intervention, these inequalities can indeed be greatly alleviated within a viable capitalist framework; without, that is, seriously undermining either the freedom or the productivity that constitute capitalism's strengths. In this chapter and the next, I shall try to show exactly how.

In this chapter, the main focus shall be on equal access to those goods and services necessary for a decent life. Or, to be more exact, it shall be on how to eliminate most poverty, for an elimination of poverty entails equal access to these necessities, even though equal access to these necessities does not entail the elimination of poverty (i.e. people could all be equally poor). In the next chapter, the main focus shall be on equality of opportunity. But since equality of opportunity and the elimination of poverty are so closely related—other things being equal, greater realization of the one automatically constitutes greater realization of the other—these two chapters complement one another.

Any overall, governmental policy for alleviating poverty should have three basic, interrelated components: a *growth* component, a *redistributive* component, and a *jobs* component. The growth component attacks poverty through stimulating economic growth, which serves to raise everyone's standard of living. The redistributive component attacks poverty through redistributing a certain amount of wealth from the rich to the poor. The jobs component attacks poverty through creating job opportunities for those whose poverty is a consequence of unemployment.

All three components are necessary. Economic growth is necessary, since it attacks the disease, so to speak, rather than merely treating the symptoms. This growth, incidentally, must be within the parameters of a clean and healthy environment. To the extent that a clean and healthy environment is incompatible with growth, the environment must be given priority. Most economists believe, however, that, even within these environmental parameters, substantial growth is still possible. Economic growth means higher productivity per worker, which, in turn, means higher wages per worker and therefore less poverty. How to achieve economic growth is no secret: it is brought about through capital accumulation. And here I am referring to the accumulation of not only 'physical' capital—factories, machines, and so on—but also 'human' capital—well-trained human beings—and knowledge capital—the store of accumulated technical know-how. Capital accumulation, in turn, can only be achieved by postponing some consumption or, in other words, by increasing savings and investment. An increase in savings and investment can be brought about in a number of ways. Two of the most important of these are, first, converting the government's primary means of raising revenue from a tax on income to a tax on consumption and, secondly, prohibiting governmental deficit spending (but in a way that is more effective and less likely to destabilize the economy than the US's Gramm–Rudman–Hollings balanced budget statute; see Seidman 1990: ch. 4).

Yet, as important as economic growth is, focusing solely upon it to reduce poverty has serious drawbacks. First, even employing the best of means, economic growth is relatively slow and somewhat uncertain, whereas people need relief from poverty now. Secondly, although growth may succeed in raising the absolute position of the poor so that they no longer fall below minimal subsistence, it may do little or nothing to improve their relative position—that is, even if economic growth has succeeded in raising everyone's absolute position, the least well off may well remain, relative to those better off, just as bad off, with therefore just as little in the way of opportunities to succeed, as ever. Thus, any satisfactory antipoverty policy must have a redistributive component as well.

Finally, poverty today is at least as much a result of unemployment as it is of low wages. A society may be able to afford enough redistribution to remedy low wages, but no society can possibly afford enough redistribution permanently to support, in full, a large number of 'able-bodied' people who are unemployed. Not even economic growth is necessarily a remedy for unemployment, for economic growth can be achieved largely through the greater productivity of those already employed, with relatively little gain in employment. And, in any case, economic growth is a slow process. So, in addition to growth and redistributive components, any satisfactory, overall antipoverty policy must include a jobs component as well.

Philosophers have often enough argued that the poverty amidst plenty found in capitalist countries cannot be morally justified. Defenders of the status quo generally try to circumvent these arguments by claiming that this poverty is, for the most part, necessary—necessary, that is, if we are to continue to enjoy capitalism's considerable advantages in the areas of freedom and productivity. Any governmental programmes extensive enough to eliminate most poverty, they claim, will undermine productivity largely by undermining people's incentives to work. Since most people seem to find these claims persuasive, further philosophical arguments based solely upon the immorality of poverty are therefore probably futile. What is needed instead is a convincing *practical* argument, one designed to show that a governmental antipoverty policy need not have the dire consequences upon freedom and productivity that defenders of the status quo claim. But I know of no way to present such an argument short of actually setting out, in some detail, just such a policy—that is, an antipoverty policy that does not seriously undermine either productivity or freedom. What follows is therefore an attempt to set out a policy that eliminates most poverty and does so without serious losses in work incentives, overall productivity, or freedom.

The antipoverty policy I shall set out and defend here consists of the three basic, interrelated components identified above. Two of these three

components—the redistributive component and the jobs component—
shall be set out in some detail here. The third component—growth—is
equally important, but shall be set out in less detail here. This component
consists of ways to increase a nation's capital—physical capital, human
capital, and knowledge capital. The main way of increasing human and, to
some extent, knowledge capital to be proposed here is by increasing
educational benefits. And the importance of human and knowledge capital
for economic growth ought not to be underestimated (see e.g. Reich 1989:
ch. 17). But these same educational benefits are also part of the
redistributive and jobs components of the overall policy to be set out here,
and thus need not be considered separately as part of the growth
component. Ways for increasing *physical* capital, on the other hand,
involve technicalities largely beyond the scope of this enquiry.

5.2 ANTIPOVERTY POLICY, 1990

Let us begin by looking briefly at the poverty found in a typical capitalist
country. Consider the United States in 1990. In 1990 no less than one out
of every ten US citizens, and one out of every five children, lived in
poverty.

What shall be meant here by 'living in poverty' is living in a family whose
total (pre-tax) income is below the official United States government
poverty line, a line that is supposed to represent the minimum income
necessary for subsistence. The government determines what counts as a
minimum subsistence income by taking the cost of a minimal subsistence
diet, and multiplying by three (since, typically, families spend from one-
third to one-quarter of their total income on food). Whether this is the
most appropriate way to measure poverty is controversial (see e.g.
Chambers 1981). But this is a controversy that shall not be addressed here.
Throughout this discussion only the official 'weighted average' poverty
lines, the ones most often quoted, shall be used. Alternative ways of
measuring poverty might require some changes in how the antipoverty
policy to be set out here is illustrated, but would not require any changes in
the policy itself.

Children living in poverty are, typically, malnourished, diseased, ignorant,
susceptible to drugs, and rejected by others. As a result, they are often left
with permanent physical or psychological handicaps that make it virtually
certain they shall never realize their potential in life as other children do.
Surely, if for no other reason than this great injustice and tragic loss of
human potential, this amount of poverty in a relatively rich country is
intolerable.

Let us now consider US antipoverty policy in 1990, since it serves as an instructive bench-mark—an excellent example of bad policy—with which better policies can be usefully compared. And as I write, incidentally, most of 1990 policy, as bad as it was, remains intact. Although I am thus using 1990, *United States* antipoverty with which to compare the alternative policy to be set out here, the general arguments in favour of this alternative apply to any modern, industrialized country.

First, a few matters of terminology. Governmental transfer programmes may be classified as being either poverty-related, or not. By a 'poverty-related' programme, I mean one that distributes benefits to people only if they can demonstrate that they are living in (or near) poverty. Poverty-related programmes are often referred to as ones that are 'means-tested'. I prefer, however, to reserve the term 'means-tested' for those programmes the benefits from which vary as a function of the recipients' means—that are, in other words, '*income*-related'. Thus, as I am using these terms, all poverty-related programmes are means-tested, but not all means-tested programmes are poverty-related, since means-tested programmes may provide some benefits to the middle and upper classes as well.

Next, governmental transfer programmes—both poverty-related ones and ones that are not—may provide either 'in-kind' benefits or 'cash' benefits. In-kind benefits are either goods or services of some sort, such as food or education, or else something, such as coupons or vouchers, redeemable only for certain goods or services.

Finally, not all aspects of an overall, antipoverty 'policy' need, of course, take the form of a transfer programme. In fact, minimum-wage legislation, one of the main features of 1990 US policy, does not take this form, but is as follows.

5.2.1 Minimum-Wage Legislation

The simple idea behind minimum-wage legislation, so central to 1990 US antipoverty policy, is, of course, that by forcing employers to pay employees more, there will be less poverty. Unfortunately, the solution to the poverty problem is not quite this simple; minimum-wage legislation very likely does as much harm as good. Although minimum wages help counter one of the two main causes of poverty—namely, low wages—they contribute to the other main cause of poverty—namely, unemployment. With minimum wages, many marginal jobs, whose market value is below the minimum wage, are lost. Exactly how many jobs are thus lost is controversial, but it is clear that the number is significant. Furthermore, marginal firms, ones that can only afford to remain in business by paying less than minimum wages, are, with minimum-wage legislation, forced out of business, resulting in still more lost jobs. And these lost jobs and firms add up to an overall loss in productivity throughout the entire economy, not to mention an increased burden from the greater unemployment compensation and welfare payments to which these lost jobs give rise. Also, to the extent that firms do manage to raise previously low wages up to the minimum wage, they will typically pass this expense on to the public in the form of higher prices. Thus a certain amount of what low-income workers gain through minimum-wage legislation, they may end up losing through having to pay more for their goods and services.

Finally, minimum wages are an inefficient way of reducing poverty for still another reason: much of the increase in wages mandated by minimum-wage legislation does not go to those in genuinely poor families, but goes instead to low-wage earners in affluent families, such as teenage children with summer jobs.

5.2.2 Non-Poverty-Related Programmes

Let us turn now to governmental transfer programmes. Consider, first, ones that do not limit their benefits exclusively to those who are poor or near-poor. They are not, in other words, 'poverty-related' programmes.

The most important 1990 non-poverty-related programme (aside from public education) was Social Security, which provided retirement benefits for everyone. Other such programmes were disability insurance and unemployment insurance. Since eligibility for benefits from these pro-grammes did not depend upon being poor or near poor, they were generally not viewed as antipoverty programmes. But their role in reducing poverty qualifies them as being integral parts of 1990 antipoverty policy nevertheless. These programmes, although far from perfect, were among the best features of 1990 antipoverty policy.

5.2.3 Poverty-Related Programmes: In-Kind Benefits

Let us turn next to poverty-related programmes, beginning with one that provided in-kind benefits. Perhaps the prototype of such programmes was the Food Stamp programme, jointly operated by federal and state governments. Under this programme, families that demonstrate sufficient need receive monthly food stamps that may be used to purchase food at any retail stores. Other in-kind programmes included rent supplements for the poor and public housing—public housing being a programme that called for housing units to be built by the government and rented to the poor at subsidized rates.

In-kind programmes do indeed help the poor. But they do not do it in the best way. First, in-kind programmes are unnecessarily paternalistic. By means of these programmes the government, in effect, says to the poor: 'You are not capable of spending cash in a way that is in your best interests;

therefore we will provide you with benefits in the form of food and housing instead.' Aside from the insult, this unnecessary paternalism has the following disadvantage. With X amount of benefits in the form of food stamps and rent supplements rather than cash, a poor family must purchase X amount of food and housing even if, were it given X in cash instead, it could purchase adequate food and housing for less than X, thus leaving it cash with which to purchase still other necessities. In other words, in-kind benefits preclude the most efficient use of resources.

The counter-argument, of course, is that the poor are indeed not capable of spending cash in a way that is in their best interests, but will spend it on liquor, tobacco, and recreation instead; thus, for their own good (and that of their children), benefits must take the form of food and shelter. It has been found, however, that families living in poverty without any public support voluntarily devote 63.8 per cent of their incomes to food and housing anyway, compared to 53.8 per cent by middle-income families (Browning 1975: 47–8). And these same poor families spend only 5.4 per cent of their incomes on liquor, tobacco, and recreation, compared to 7.3 per cent by middle-income families. There are, to be sure, always some irresponsible people in every income group, but, as this study suggests, the majority of poor are responsible, and it is inappropriate to treat them all as if they were like the minority who are not.

A second disadvantage of in-kind programmes is that they are far more expensive to administer than cash programmes such as the earned income credit (see below), which can easily be administered by the Internal Revenue Service without creating a new bureaucracy (Browning 1975: 50). Similarly, with a complex system of in-kind benefits, welfare fraud is harder to control than with a simple system of cash benefits.

But perhaps the most serious disadvantage of poverty-related, in-kind programmes is this: they are demeaning to the poor, setting them off from the rest of society. One's acceptance of in-kind benefits is usually quite obvious; it is hard to hide the fact that one lives in public housing, or that one pays for food with food stamps. In a capitalist society, self-respect and respect from others is largely a function of one's income. This may well not be a good thing but, whether a good thing or not, it is a fact unlikely to change. So accepting in-kind benefits says to all the world: 'I am a poor person and, as such, unworthy of much respect.'

Moreover, because in-kind benefits are in fact demeaning and also because the various applications for these benefits are, typically, complex and confusing, many of the poor, who are eligible for these benefits and desperately need them, do not even apply for them.

Because of these disadvantages, in-kind benefits are generally less satisfactory than cash benefits. This does not mean that no type of in-kind benefits should be part of an overall antipoverty policy. On the contrary, at least two types of in-kind benefits are crucial: medical and educational benefits. This is because, unlike food and housing expenses, which are relatively moderate, invariable, and predictable, medical and educational expenses are neither invariable nor predictable, and may well not be moderate. They are, in other words, 'extraordinary' expenses and, as such, the means for handling them cannot very well be provided through any programme of regular cash benefits. But more on this later (Sect. 5.6).

5.2.4 *Poverty-Related Programmes: Cash Benefits*

Let us turn, finally, to the poverty-related, cash-benefit programmes of 1990, US antipoverty policy. By far the most costly and important of these programmes was Aid to Families with Dependent Children (AFDC). AFDC is, by its terms, a joint federal and state venture. The federal government sets the broad guidelines for the programme and supplies much of the funds. Each of the fifty US states, within these broad federal guidelines, set their own more specific requirements for eligibility and are charged with administering the programme.

Unfortunately, the AFDC programme was poorly conceived. First, it was inconsistent. Because of state autonomy in working out the details, eligibility requirements (along with the amount of benefits) varied greatly from state to state. In 1990 some states provided almost four times the benefits of other states. Is it fair for poor people to receive radically different treatment depending merely upon where they happen to live within the country?

Secondly, benefits were inadequate in amount; no state as of 1990 provided enough benefits to keep an AFDC family out of poverty; most states did not even come close. Moreover, with AFDC each state has good reason to keep its benefits inadequate; if a state's benefits were adequate, then the poor from other states would have good reason to move to that state just to take advantage of these benefits, and no state wants the burden of caring for the poor from other states.

Thirdly, AFDC left large numbers of poor families with few or no benefits at all. In some states, eligibility for full, continuing AFDC benefits depended upon the family being headed by a single parent, thereby providing no benefits at all for poor families headed by two parents. Moreover, as soon as those already receiving benefits got a job, AFDC benefits were greatly reduced and cut off altogether after a year, no matter how inadequate the family's income remained.

Fourthly, those states in which full, continuing benefits depend upon the family being headed by a single parent provided a perverse incentive for parents not to marry. The extent to which AFDC actually did encourage single-parent households is controversial (Wilson 1987: chs. 3 and 4;

Duncan and Hoffman 1988: 249–50; Kaus 1992: ch. 7). In any case, that married parents not be treated less favourably than single parents is simply a requirement of elementary justice.

Fifthly, by abruptly cutting off all AFDC benefits, along with other benefits that are 'tied' to AFDC, one year after gaining employment, the programme acted as a disincentive to work—unless, that is, one were paid substantially more than one could receive in AFDC and related benefits.

In sum: intelligent people of good will ought to be able to devise a programme that is more satisfactory than AFDC.

Actually, 1990 policy did include a more satisfactory cash-benefit programme: the earned income credit (EIC). The basic idea of EIC is simple: those with very low income receive a tax credit equal in amount to a certain percentage of all the income they earn up to a certain cut-off point. This credit is used to offset any Federal income taxes they owe. More importantly, to the extent that the taxes they owe are less than their credit, the government sends them a refund check for the difference. This refund thus acts as a wage supplement. To be eligible for this credit, one must have at least one child who qualifies as a dependent. Finally, as one's earned income continues to increase beyond the cut-off point, the credit is gradually phased out.

The main objection to EIC is that the programme was far too modest. A maximum benefit of $910 in 1990 was hardly enough to raise many families above the poverty line. I shall have more to say about EIC later.

This completes my brief sketch of 1990 antipoverty policy. Although we encountered a few bright spots, the overall picture that emerges is that of a policy which was inconsistent, provided inadequate benefits, was administratively inefficient, gave rise to perverse incentives, failed to create job opportunities, and was unnecessarily paternalistic and demeaning.

These objectionable features have led many conservatives to conclude that, therefore, government antipoverty policies should be eliminated altogether (e.g. Murray 1984). But these conservatives have things backwards. What should be eliminated are not government antipoverty policies, but these objectionable features. How is this possible? I shall try to show how next.

5.3 THE PROPOSED POLICY: AN OVERVIEW

What follows is an antipoverty policy that, I shall argue, avoids the objectionable features of 1990 policy. More generally, it is a policy that eliminates most poverty without any serious loss in overall productivity or personal freedom. In this section, this policy shall be briefly outlined and, in subsequent sections, it shall be examined in more detail.

For purposes of this policy, poor people who are 'heads of households' are to be categorized into two groups: (I) those who work but, nevertheless, do not earn enough to keep them or their families above the poverty line, and (II) those who do not work. Those who do not work are, in turn, to be categorized into five further groups: (A) those who do not work because they have reached retirement age and have retired, (B) those who do not work because they are mentally or physically disabled, (C) those who do not work because they cannot find a job, (D) those who do not work because they are single parents who must take care of young children, and (E) those who do not work simply because they do not want to work. There is, of course, some overlap among some of the above five groups.

For those who fall into categories II.A and II.B, the policy being set out here simply incorporates, by and large, any current programmes that provide annuities for the retired (i.e. Social Security) and the disabled (i.e. disability insurance). These state-financed 'insurance' programmes may be far from perfect. Yet without them many of the poorer people throughout society would remain unprotected against disability and old age. These people could hardly be expected to pay, to private companies, the premiums for this insurance if, as would all too often be the case, paying these premiums meant doing without adequate food on the table or clothes for their children (see Sect. 2.5). Thus these state-financed insurance programmes should either be retained (with certain modifications, such as suggested by Haveman 1988: 153–63), or else replaced by other programmes in the form of government vouchers earmarked for this insurance.

For those who fall into category II.C—those who do not work because they cannot find a job—the policy being set out here takes a threefold approach. First, state-financed unemployment insurance is to be retained or improved. Secondly, educational and job training opportunities are to be made far more readily available. And thirdly, certain measures are to be taken to assure the existence of a job for anyone willing to work. Of course assuring the existence of a job for anyone willing to work is easier said than done, but certain things, beyond what the government already does, can indeed be done. The first thing to be done, according to the policy being set out here, is to abolish minimum-wage legislation. How many new jobs this will create is controversial, but most economists believe the number is substantial. Other things are to be done as well, but more on this later.

For those in category II.D, those who do not work because they are single parents with young children to take care of, the policy being set out here calls for the establishment of an adequate, comprehensive system of child-care centres and related programmes. But more on this later also.

For those in category II.E, those who do not work simply because they do not want to work, the policy being set out here does nothing—that is, nothing beyond merely providing the medical and educational benefits to which everyone is to be entitled. I know of no way of providing full, long-term governmental benefits to those currently falling into this category without the number of people falling into this category immediately expanding to an unacceptable amount. (To this extent, Murray (1984) is correct.) No society—whether capitalist or socialist—can afford to subsidize, indefinitely and at a comfortable level, those who do not work simply because they do not want to work, and the policy set out here does not do so. Just as the conservative says: doing so would undermine other people's incentives to work. This does not mean people who do not want to work will be left to starve. According to the policy being set out here, people falling into this category are best taken care of through the soup kitchens, old clothes, and so on, of local programmes and private efforts.

Let it be noted, incidentally, that, in the United States, not all those who simply do not want to work are merely lazy. Many poor, urban blacks and other minorities do not want to work because, after generations of vicious discrimination, unemployment, and poverty, they no longer have any faith in education, employment, and the so-called American dream. These people are not lazy; rather, they are completely alienated from the system. They have dropped out of the labour market altogether (Kirkland 1992). They are the 'truly disadvantaged' (Wilson 1987). How to reverse their self-destructive attitude is a difficult and important question. Eliminating discrimination is, of course, a necessary and, perhaps, sufficient condition for doing so in the long run, but what about the shorter run? I do not think that the answer is preferential treatment, for, as Wilson points out (1987: chs. 5 and 9), to benefit from preferential treatment, one must already have a positive attitude towards work; that is, one must already have escaped from the underclass of truly disadvantaged. Nor do I think the answer is in terms of a 'free ride' from governmental welfare, which will likely change few attitudes while bleeding the country dry. Perhaps the answer to how these self-destructive attitudes can be changed lies largely, once again, with local programmes and private efforts. But what also is needed is a just, humane antipoverty policy, such as I am trying to set out here, for if these self-destructive attitudes ever do begin to change through local and private efforts, then, so as to solidify any such changes, it is essential that the appropriate programmes and opportunities for benefiting from positive changes already be in place. And if, by means of these programmes and opportunities, the system, for these people, finally will be just and humane, then will not this very fact, by itself, be enough to cause many to be less alienated?

We come, finally, to those in category I, those who are poor even though working. In 1990 close to 50 per cent of the principle wage-earners who were poor in the United States worked, with about one-third of these working year-round, full-time. It is a shame that, with 1990 antipoverty policy, the working poor, perhaps those most worthy of support, received the least. Surely people who—usually through no fault of their own— simply do not have marketable capacities scarce enough to keep them from being poor in an economy where income is determined by supply and demand, yet who do their fair share of work anyway, deserve society's support. What I am saying, in other words, is this. Those who enjoy the considerable benefits of a free-enterprise system where incomes are determined by supply and demand ought, morally speaking, to take care of the system's innocent victims—those who work as hard as anyone else, and usually at jobs far more miserable than most, but who simply do not command the earning power to keep them from poverty.

Indeed, I know of no major approach to morality, except one, that does not require adequate social support for these victims of the system. Rawls's difference principle—which prescribes that we make the worst off as well off as possible—requires it; Kant's categorical imperative—which prescribes that we treat no one as a mere means—requires it; direct utilitarianism, in the light of the diminishing marginal utility of wealth, requires it.

And indirect utilitarianism, the approach defended here (Ch. 1 and Sect. 2.2), also requires support for these victims of the system. Indirect utilitarianism, as we have seen, stands for certain norms and values. Among these norms are certain rights, such as free speech. Among these values are equal access for everyone to necessities and equal opportunity. For working people to suffer from poverty in a nation of plenty is— obviously—incompatible with the values of equal access for everyone to necessities and equal opportunity. Moreover (as can be seen from Ch. 2), the measures set out here for relieving this poverty do not violate any indirect utilitarian rights. So the only question is whether these measures, urgently called for by the above two values, are nevertheless incompatible with other indirect utilitarian values—especially those of freedom and productivity—thereby creating a serious conflict of values. If so, then the next step would be an appeal to the 'general welfare' for adjudicating this conflict and arriving at the best possible compromise (Sect. 2.2). But these measures for relieving poverty are, as we shall see, especially designed to interfere as little as possible with the market and with people's work incentives. And, as we shall also see, the opportunities that are created by these measures—educational opportunities, job opportunities, growth opportunities—are far more likely, in the long run, to increase overall freedom and productivity than decrease them. It is doubtful, therefore, that these measures will give rise to any serious conflict that necessitates an appeal to the general welfare. If, however, it did come down to an appeal to the general welfare, then direct utilitarian considerations, such as, once again, the diminishing marginal utility of wealth, would tip the balance in favour of the working poor anyway.

The only major approach I know of that does not require support for these victims of the system is libertarianism. But since libertarianism rules out virtually any governmental activity of any sort, it is hard to take it seriously (see Ch. 2, however, where the libertarian approach is in fact taken seriously and rejected).

The problem of how to provide adequate support for the working poor is especially pressing if, as is advocated here, minimum wages are to be abolished.

The solution set out here is, very briefly, as follows. First, current antipoverty policy (with the exceptions already noted) is to be abolished. For the USA in 1990, this would have meant abolishing minimum wages, AFDC, all 1990 tax exemptions, the 1990 earned income credit, Medicaid, Medicare, food stamps, housing subsidies for the poor, and numerous less important programmes. In place of what is abolished there is to be (1) a new, more meaningful tax exemption, (2) a new, far more adequate earned income credit, and (3) non-poverty-related medical and educational benefits to help cover the major extraordinary expenses with which families are likely to be confronted.

This then completes the overall sketch of the antipoverty policy being set out here; it is now time to turn to the details. I shall, for simplicity, be assuming throughout that the government is taxing income rather than consumption; the overall policy being set out here, however, can be tailored to the requirements of a consumption tax as well. And I shall be illustrating this policy by showing how it would have applied to the United States in 1990. But, once again, the basic arguments for something like this policy are meant to be equally applicable to any modern industrialized country at any time.

5.4 THE STANDARD EXEMPTION

The antipoverty policy set out here presupposes one thing: the government should not, through taxes, take income away from a household below the poverty line, a line that represents the *minimum* income required for subsistence. When the amount necessary for financing public spending can easily be raised from those who are comfortably well off or rich, it is surely a mean people that demands a tax contribution from those already below the poverty line, thus forcing them even deeper into poverty. Even if the richer people of a nation are selfish and greedy enough not to make those among them suffering in poverty any better off, they should at least refrain from deliberately making them worse off, from deliberately taking the very necessities of life from them. Surely this is the very least that can be expected from any humane antipoverty policy.

Accordingly, the first aspect of the antipoverty policy being set out here is this: all people are to receive an exemption from any sort of national taxes on their income up to an amount that equals the poverty line for a single person without children. And so as to leave the government no loopholes, this exemption is to apply not just to ordinary income taxes, but to any such abominations of US FICA income taxes as well. Currently, FICA income taxes—that is, Federal Insurance Contributions Act taxes— are shielded from exemptions and deductions.

Everyone, not just the poor, is to be entitled to this standard exemption so as to avoid the notorious 'notch' problem. The notch problem exists whenever, with an increase in one's (pre-tax) income, one's taxes then go up so much, or one's governmental benefits go down so much, that it more or less wipes out this increase. Any such notches inherent in tax law serve as serious disincentives for improving one's situation. With everyone being entitled to the standard exemption, no one will, upon crossing a certain income barrier, immediately lose this exemption, and thus suffer a sudden and dramatic increase in income-tax liability.

An alternative way of avoiding the notch would be to phase out the standard exemption gradually for all those over a certain income. A gradual phasing out of this exemption is a perfectly reasonable way for the government to increase tax revenue. But I shall not pursue this particular complication.

The standard exemption is to be determined each year by reference to the poverty-line figures for the previous year, since these are always the latest figures available. To prevent the standard exemption from changing too often from year to year as inflation drives the poverty line up, the exact amount of this exemption could perhaps be determined by rounding off the poverty line to the nearest X, where X equalled some figure large enough to prevent the exemption from changing more than, say, once every five or six years (and thus rounding off would also have the advantage of simplifying the mathematics). In the USA, for example, let us assume that X equalled $1,000. Thus, in the USA, the standard exemption would equal the previous year's poverty line rounded off to the nearest $1,000. Since the poverty line for a US citizen with no dependents in 1989 equalled $6,311 (US Bureau of the Census 1990: 108), the standard exemption for 1990 would then have been $6,000.

In the USA, at least, this exemption from all national income taxes is probably best accomplished by abolishing FICA income taxes—that is, social security taxes—thus raising the money for Social Security, and related programmes, through ordinary income taxes. This means abolishing the FICA contributions made by employers as well, since these contributions probably end up being borne mostly by employees too, through lower wages than otherwise. Why not, indeed, change FICA taxes into ordinary taxes? Revenue from FICA taxes is being used for purposes other than Social Security and related programmes anyway, just like ordinary taxes. Some argue against simplifying matters by collapsing these two taxes into one by claiming that keeping FICA taxes separate preserves the illusion that what Americans pay in support of Social Security and related programmes are insurance premiums, not taxes. But why try to preserve this illusion? Perhaps the idea is that this illusion helps assure continued public support for Social Security and related programmes. This idea, however, rests upon a mistaken premiss. The public's continued support of these programmes has nothing to do with their being thought of as insurance, but is instead a consequence of their being universal, rather than poverty-related, programmes. As universal programmes, the benefits they provide are not stigmatized as 'welfare', and therefore are acceptable to rich and poor alike. Continuing to separate FICA income taxes, which are regressive, from ordinary income taxes serves only to make it easier for politicians to extract taxes from the poor. This shameful farce should be discontinued.

The standard exemption proposed here—as modest a proposal as it is—should at least be sufficient for preventing most single people without children from falling below the poverty line—assuming, that is, that the jobs component of the overall antipoverty programme works. Even without minimum wages, it would have been unlikely for an American to work full-time throughout the year in 1990, yet earn less than what would have been the 1990 standard exemption of $6,000.

But then what about those without children who are not single but married? After all, the 1989 poverty line for a married couple without children was $8,076, substantially higher than this $6,000 exemption. The answer is as follows. In the case where the husband, wife, or both suffer from a disability that prevents work, the couple will be eligible for disability benefits of an amount that is, or should be made, sufficient to keep them above the poverty line. And in the case where both are healthy they should have little problem staying above the poverty line either since, if necessary, both can work—again, assuming the availability of jobs. If both worked, their combined post-tax income would probably place them well above the poverty line. This is especially true since, although they are to be taxed as a unit upon their combined incomes, both are to retain the standard exemption that would be available to him or her if single. The way this is to work is that, just as would be the case if both were single, each is to use his or her exemption only to the extent that he or she earns income. Thus, given a $6,000 standard exemption, a married couple, one of whom has $12,000 income, the other of whom has $3,000 income, will have a $9,000 exemption between them (the one earning $12,000 accounts for $6,000 of this $9,000, and the one earning $3,000 accounts for $3,000 of it).

Allowing married couples to retain the exemption that would be available to each if single accomplishes two things. First, this makes it very likely indeed that working couples without children, even with both earning low income, will nevertheless remain above the poverty line after taxes. Secondly, and even more important, is this. Both fairness and the goal of not undermining family values demand as much tax-neutrality as possible as between being single and being married; and allowing married people to have the same exemption as they would have if single helps achieve this neutrality. With tax neutrality (which, incidentally, 1990 US tax law was far from achieving) decisions as to whether to get married or not can then be based upon legitimate reasons rather than tax considerations.

5.5 THE REFUNDABLE CREDIT

As pointed out above, even after minimum wages are abolished, most people working full-time throughout the year can be expected to earn at least enough to keep a single person without children above the poverty line. Thus the standard exemption (Sect. 5.4), along with protection against extraordinary expenses (Sect. 5.6), should be a sufficient remedy against poverty for most working people without children. It is time now to examine what should be done about the poor, either single or married, with children. To stay out of poverty, people with children obviously require more income than those without; the 1989 poverty line for a single person with one child was $8,076, for a single person with two children, $9,885, and for a single person with three children, $12,675, approximately double that which it was for a single person without any children. (Remember: for 1990, it would have been 1989 poverty-line figures, the latest ones available then, that would have been relevant.)

An obvious possibility for helping the poor with children would be simply to increase their standard exemption to an amount equal to the poverty line for them. But this will not work. In order to benefit fully from a larger exemption, one must first have income equal to this larger exemption, and that is the problem. If minimum wages are abolished as is advocated here, many poor, although working full-time throughout the year, may not earn much more than an amount equal to the poverty line for a single person without children; thus to increase their standard exemption above this amount would be of little use to them.

Instead of a larger standard exemption, the best way to alleviate poverty among those with children is neither new nor radical; in fact, something very similar was already part of 1990 US tax law. The best way to alleviate poverty is through a version of the earned income credit (EIC). As will be recalled, the 1990 EIC provides a credit against taxes for those with dependent children. The amount of this credit is a percentage of the amount of income, up to a certain point, that one earns. The key feature of this credit is that, as opposed to a normal tax exemption, it is refundable, in that one receives cash from the government for any part of one's credit that cannot be used to offset one's taxes. Indeed, that the holder of a credit receive a cash refund from the government for any part of it that cannot be used to offset taxes is only fair; without this cash refund, those with lower incomes would get little or no benefit at all from the credit, yet they are the very ones who need it the most. Because the amount of this credit is proportional to one's wages, and because it is refundable, it is a kind of wage supplement. Finally, as one's income increases beyond a certain point, and one's need for a wage supplement thus decreases, the refundable credit is gradually phased out.

But what is being proposed here differs from the 1990 EIC in the following two important ways. First, contrary to the 1990 EIC, the amount of credit one gets is to increase with the number of dependent children one has—up to three children. So the greater one's child expenses, the higher one's credit will generally be.

To continue increasing the amount of this credit beyond three children might, however, be a mistake. Doing so might encourage those who can least afford it to have large numbers of children. And increasing the amount beyond three children might increase the costs to taxpayers beyond what could be afforded, especially if doing so did in fact encourage the poor to have more children. So, with what is being proposed here, if benefits beyond what is available with three children become necessary for those who insist upon having more than three children, they must then rely upon local or private charities. And existing laws that provide for removing children from unfit homes and placing them elsewhere will have to be the remedy of last resort.

As mentioned earlier, the maximum credit available with the 1990 EIC was a mere $910, hardly enough to make much difference in the amount in poverty in the United States, considering the fact that (for example) the poverty line for a single-parent family with two children was close to $10,000. This gives rise to the second, and most important, way in which what is proposed here differs from the 1990 EIC. The refundable credit proposed here is to be large enough to really eliminate poverty. The general idea is for it to be large enough to eliminate virtually all poverty for working families with up to three dependent children. We may assume that the income of virtually any working family will at least equal the standard exemption (that is, their income will at least equal the poverty line for a single person with no dependents). Therefore, the refundable credit proposed here is to be large enough so that, when added to income equal in amount to the standard exemption, the resulting sum will, for any single-parent family with up to three children, closely approximate their poverty line.

But, it might now be objected, this refundable credit is to be large enough only to eliminate virtually all poverty for working families with up to three dependent children provided that they are single-parent families. What about poor *two-parent* families to which a somewhat higher poverty line is applicable? The answer is that, in two-parent families, both husband and wife can work if necessary (remember: inexpensive child care is to be available to all, and other measures, explained below, are to be taken to assure that those who want to work will be able to do so). The difference between the poverty lines for single-parent and two-parent families is not great. With two children, for example, the difference was only $9,885 versus $12,675. Two incomes should, therefore, be more than sufficient to cover the higher poverty lines applicable to two-parent families—especially since, as we saw, both husband and wife are to be entitled to their own standard exemption.

As family income increases, the refundable credit is to be gradually phased out. This phase out is to begin with income greater than the standard exemption, and proceed at a rate designed to eliminate the credit entirely for families with incomes approximately equal to average family income throughout the country. This, of course, means that not just very poor families, but all families with dependent children and an income less than average will enjoy at least some bonus income from the EIC, the less their income, the greater their bonus. And given the expense of children, it should be a much needed bonus indeed.

The amount of one's credit that is phased out is to be determined not merely on the basis of one's 'earned' income, as is one's credit, but on the basis of one's income from any source. This is so as to prevent those who have low earned income but high unearned income (i.e. from stocks, bonds, and so on), and who therefore presumably do not need this credit, from receiving it anyway.

Finally, the amount of credit that a two-parent family is eligible to receive is to be determined by the combined earnings of both husband and wife; their phase out, however, is to begin when their combined incomes reach an amount equal to a single standard exemption, even though, for determining the couple's overall tax liability, they may, as explained previously, be entitled to a combined exemption equal to as much as twice the amount of a single standard exemption. Making the amount of the credit depend upon the couple's combined incomes acts as a slight disincentive for a working, single parent to get married. On the other hand, any such disincentive is offset by the economies of living as a married couple, and by the fact that a father will have an incentive to marry the mother of his children (assuming she is the one with custody) for only by doing so will *he* become eligible for the credit.

Let me try to make this EIC programme more clear with an illustration of its being put into practice. Take, once again, the United States in the year 1990, and consider, first, a single-parent family with one child. To bring such a family with an income equal to what would have been the standard 1990 exemption of $6,000 up to approximately their poverty line of $8,076 (as called for by this programme), a credit of about 33⅓ per cent of income would have been necessary. This means, of course, a credit of 33⅓ per cent of income up to a cut-off point of income equal in amount to the standard exemption. Thirty-three and one-third per cent of their $6,000 income would have given this family a credit of $2,000. Since their income of $6,000 would not have exceeded the standard exemption, they would have owed no federal income taxes with which to offset this credit; therefore, it would have taken the form of a $2,000 cash refund from the government. This $2,000 cash refund, added to their $6,000 income, would then have resulted in a sum of $8,000, which would indeed have approximated their poverty line of $8,076. To bring single-parent families earning $6,000, but with two children, approximately up to their poverty line of $9,885, a credit of about 66⅔ per cent of their income would have been necessary. Finally, to bring single-parent families earning $6,000, but with three children, approximately up to their poverty line of $12,675, a credit of about 100 per cent of their income would have been necessary. The percentages of income available as credits can, of course, always be set at amounts so as to hit the relevant poverty lines exactly but, so as to make the mathematics of this illustration as simple as possible, let us assume that, in 1990, the above percentages of 33⅓, 66⅔, and 100 would indeed have been chosen.

Since the standard exemption would have been $6,000, the phase out would have begun with income greater than $6,000. And since average family income was $36,520, the phase out should have ended with income around that amount. Thus, for one-child families, a phase-out rate of 6⅓ per cent of every dollar of family income over the standard exemption would have been about right; for two-child families, a phase-out rate of 13⅓ per cent; and for three (or more)-child families, a phase-out rate of 20 per cent. Given these phase-out rates, with one child the credit would have been reduced by $67 for every $1,000 of family income over $6,000; with two children, it would have been reduced by $133; and with three or more children, by $200. Thus, for all families, the credit would have been phased out entirely at an income of $36,000, or slightly below the average of $36,520.

Let us assume, merely for purposes of this illustration, a flat income tax of 33⅓ per cent (more on this later). Given this income-tax rate, and assuming, for simplicity, only one wage-earner, Table 5.1 shows the post-tax incomes that would have resulted from this 1990 refundable credit for a family with one child, a family with two children, and a family with three or more children, at various pre-tax income levels. (For interpreting Table 5.1, recall that the relevant poverty line for a single-parent, one-child family was $8,060, for a single-parent, two-child family, $9,885, and for a single-parent, three-child family, $12,675, and average family income was $36,520.)

Notice two crucial things about this EIC programme. First, it will eliminate a vast amount of poverty in such a way as will have minimal effect upon people's incentives to work. With both the AFDC programme that was current in the USA in 1990 and a widely discussed alternative programme known as a Negative Income Tax (NIT), one's welfare benefits are greatest if one has no income at all, and the more income one has, the less these benefits become (the decrease in benefits being sudden with AFDC and gradual with NIT). Thus these programmes provide an incentive not to work at all or, if working, not to increase one's income. With the EIC programme proposed here, however, if one is not working at all, there are no benefits at all. Thus, contrary to the above two programmes, this programme, obviously, does not provide an incentive not to work. Moreover, up to an income equal to the standard exemption, the greater one's income, the *greater* one's EIC benefits; thus, up to this point, there is little or no incentive to not increase one's income. And as EIC benefits are gradually phased out for families with incomes greater than the standard exemption, they are phased out gradually enough so that a family is always better off financially earning more income than they would be with less income but greater EIC benefits; in other words, these benefits are to be phased out gradually enough so as to avoid any notch problems. Therefore, any incentive to not increase one's income should, with this programme, be very minimal indeed.

To be sure, assuming a flat income tax rate of 33⅓ per cent, phase-out rates of 6⅔, 13⅔, and 20 per cent would have left those with income in the 1990 phase-out range—$6,000 to $36,000—with an effective marginal tax rate of 40 per cent if they had one child, 46⅔ per cent if they had two children, and 53⅓ per cent if they had three or more children. Their statutory marginal tax rate would have been only 33⅓ per cent like everyone else's. But since, for all practical purposes, their loss of benefits would have acted like a tax, their *effective* marginal tax rates would have been 40, 46⅔, and 53⅓ per cent. Moreover, if people's effective marginal

Table 5.1 Refundable tax credit at various income levels ($)

	(A) With one child		
Pre-tax income	Credit	Income tax	Post-tax income
3,000	1,000	0	4,000
6,000	2,000	0	8,000
9,000	1,800	1,000	9,800
12,000	1,600	2,000	11,600
18,000	1,200	4,000	15,200
24,000	800	6,000	18,800
30,000	400	8,000	22,400
36,000	0	10,000	26,000

	(B) With two children		
Pre-tax income	Credit	Income tax	Post-tax income
3,000	2,000	0	5,000
6,000	4,000	0	10,000
9,000	3,600	1,000	11,600
12,000	3,200	2,000	13,200
18,000	2,400	4,000	16,400
24,000	1,600	6,000	19,600
30,000	800	8,000	22,800
36,000	0	10,000	26,000

	(C) With three children		
Pre-tax income	Credit	Income tax	Post-tax income
3,000	3,000	0	6,000
6,000	6,000	0	12,000
9,000	5,400	1,000	13,400
12,000	4,800	2,000	14,800
18,000	3,600	4,000	17,600
24,000	2,400	6,000	20,400
30,000	1,200	8,000	23,300
36,000	0	10,000	26,000

tax rates become too high, these rates will act as a serious disincentive to their improving their income. Thus, given a 33⅓ per cent statutory marginal tax rate, a phase-out rate should probably not go much higher than the 20 per cent rate used in this illustration. But I do not think such a rate already would be too high. In the first place, the evidence from countries, such as Sweden, with high marginal rates indicates that serious disincentives do not generally kick in until marginal rates become substantially higher than 50 per cent. In the second place, the higher effective marginal tax rates for those within the phase-out range are only temporary—lasting only so long as their children remain dependents—whereas, presumably, any increase in their income will be permanent. So, no matter how high their effective marginal tax rate might be for the short run, they still have an incentive to increase their income for purposes of the long run.

The second thing to notice about this refundable tax credit programme is that it adjusts itself automatically to increases and decreases in the value of the country's currency—or, as economists like to say, it is thoroughly 'indexed' to keep up with inflation. The whole thing revolves around one crucial figure that varies as the value of the currency varies: namely, the poverty line for a single person without children. As explained earlier, this figure is used to get the standard exemption, and the standard exemption, in turn, determines the point at which this credit begins to be phased out. Moreover, the poverty lines for those with children rise or fall each year by virtually the same percentage as does the poverty line for a single person without children. For example, the poverty line for a family of four is always approximately twice that for a single person without children. This means that if the credit and the phase-out rates set out here would have been sufficient for preventing most US working families in 1990 with up to three children from falling below the poverty line applicable to them, then, regardless of changes in the value of the dollar, these same rates should be sufficient for any other year as well. Thus, once set in motion, this programme can continue in perpetuity; no cumbersome legislative adjustments to keep up with changing prices are ever needed.

5.6 EXTRAORDINARY EXPENSES

The standard exemption, along with the refundable credit, will enable most working poor to meet minimum, ordinary expenses. But what about extraordinary expenses—necessary (as opposed to luxury) expenses that are either unpredictable or irregular (i.e. they occur only a few times during one's life) and are likely to be exceedingly large. Such expenses are devastating for the poor and, without insurance, for the middle class as well. The three kinds of extraordinary expenses people are most likely to be faced with are legal fees, medical expenses, and educational expenses.

Governmental aid in meeting these expenses goes far towards bringing about equal opportunity and equal access for everyone to necessities. It is worth noting, however, that justifying governmental aid in meeting these expenses does not depend entirely upon appealing to these values (although I consider an appeal to these values to be sufficient). Any significant gains in the education and medical care of a community can be expected to have significant positive externalities—that is, side effects (not paid for) that benefit everyone. Education creates human capital and thus greater productivity, and greater productivity benefits everyone. Better medical care helps contain the spread of disease and, indirectly, benefits productivity as well. The containment of disease and greater productivity benefit, once again, everyone. As economists, even the most conservative, will tell us, positive externalities emanating from a good or service justify some sort of governmental support for it, which will serve to remedy any 'underproduction' related to these positive externalities (see Sect. 3.1 above). Or, to put it still another way, since any significant gains in the education and medical care of a community benefit not only the direct recipients of this education and medical care, but also the public as a whole, it is only reasonable for the public as a whole, through governmental aid, to help pay the necessary expenses.

5.6.1 Legal Expenses

Of the three extraordinary expenses identified above, legal expenses seem to be the ones that were provided for the most adequately in the United States as of 1990. Every US citizen has a constitutional right to an attorney if charged with a crime, and 1990 methods for providing an attorney to those unable to afford one seemed to work. For legal expenses, therefore, the antipoverty policy being set out here simply incorporates something like 1990 practice.

5.6.2 Health Care

Provision for medical expenses, however, was far from adequate in 1990. Millions of US citizens, most of whom were poor, were not protected against medical expenses by private insurance, governmental benefits, or anything else. Among those not protected were many families with children. As a result, many of these children received little or no medical attention, and some were left with permanent disabilities that could have been prevented. Furthermore, hospital and other medical expenses, and thus insurance rates, were, in 1990, climbing at an alarming rate.

The alternative defended here is a universal, comprehensive programme of income-related, cost-sharing, similar to that proposed by Martin Feldstein (1971) and Lawrence S. Seidman (1979). This programme is to be comprehensive enough to cover not only ordinary medical needs, but also special equipment or training needed for overcoming any physical or psychological handicaps. It is not, however, to cover most cosmetic surgery or other relatively unnecessary medical expenses (such as private hospital rooms in cases where semi-private rooms are medically sufficient or physical check-ups more often than recommended).

The version of this programme proposed here works as follows. Each household or family is to be required to pay 50 per cent of its medical bills out of its own pocket up to an amount equal to 10 per cent of that part of its year's income which is above the poverty line applicable to that household. This then puts a cap, equal to 10 per cent of that part of its year's income which is above the poverty line, on how much medical expenses any household has to pay in any one year. The government is to pay the remainder.

The one exception to this 50-per-cent, cost-sharing rule might be for crucial preventative measures, such as periodic physical check-ups and prenatal care, which should probably be offered 'free' to everyone; that is, entirely at taxpayer expense. Offering preventative measures 'free' will bring about greater utilization of them, and greater utilization of them will preclude not only much suffering, but perhaps even enough costly, future medical treatment for these measures to more than pay for themselves. According to one study, every $1 invested in prenatal care in the US saves $4 through reduced hospital and long-term institutional care (Rosenbaum 1983).

This health-care programme is designed to come as close as possible to making health care more or less income neutral. Health care is 'income neutral' if and only if the same amount and quality of health care entails, on the average, the same financial sacrifice for everyone, relative to his or her income. So, with income neutrality, a poor person will pay substantially less, out-of-pocket, than a rich person for the same health care, but the *relative* financial sacrifice of each will nevertheless be, more or less, the same. The more income neutral health care is, the more equal everyone's opportunity for health care is.

Other aspects of this health-care programme include a method for paying medical bills that works something like this. The patient's medical bills are to be sent to the government, with the government then paying the doctor or hospital. At the end of the year, the government is to send to each household a statement specifying the total amount of medical expenses paid by the government on behalf of all those in the household during the year. Each household is then to determine, on the basis of its income, how much of this amount it must reimburse the government, and add that amount to its federal taxes (or subtract the amount from its refund).

Moreover, this programme is to specify that the amount the government is to pay on one's behalf is to be reduced by any amounts that private insurers have paid, or will pay, on one's behalf for the same treatments. This is so as to prevent private insurers from undermining the programme's cost-sharing strategy simply by picking up the difference between one's overall medical expenses and the amount paid by the government. Finally, so as to help constrain the escalation of medical costs, physicians are to be legally forbidden from raising fees for patients who have already reached the maximum that they are required to contribute.

With the rates set out above, the very most out-of-pocket medical expenses that, say, a middle-class family with two children and an income of $30,000 would have had to pay in 1990 (given the rounded-off poverty line of $12,500 applicable to such a family) is $1,750. And those with an income at or below the poverty line would not have had to pay anything. The reason for those at or below the poverty line not having to pay anything is that, with not enough income even for bare subsistence, they would be very likely to forego essential health care for themselves and their children otherwise.

Generally speaking, for those at any level of income this programme succeeds in providing protection against extraordinary medical expenses, thus making medical expenses generally manageable for everyone. Moreover, this programme does not leave the poor, and their children, with any less than the same quality of medical care available to anyone else. This is essential for the sake of equal opportunity.

At the same time, this programme does not remove health care from the market. It preserves complete freedom to go to whatever doctor one chooses. And, aside from fees for 'free' preventative measures, it is designed so that medical fees can be left, largely, to supply and demand rather than being determined by the government. As argued in Chapter 3—except for monopolies, externalities, or emergencies, or whenever income is distributed so unequally as to keep many below the poverty line—market prices reflect true wants and needs more accurately and give rise to a more productive economy than do governmentally determined prices. Thus, here as elsewhere, I am assuming that, given the qualifications noted above, market prices are to be preferred.

The British programme of 'socialized' medicine, an alternative to the programme defended here, cannot, of course, be dismissed without an extensive discussion far beyond the scope of this enquiry. The British programme, however, does, contrary to the programme defended here, have the disadvantage of relying almost entirely upon governmentally determined prices.

The main danger of combining universal, comprehensive health-care coverage with reliance upon market prices, as does the programme defended here, is that a 'cost-is-no-object-since-my-coverage-pays-all' mentality may cause market prices to escalate out of control. The mechanism built into the programme for preventing this escalation is cost-sharing. The idea of cost-sharing is that both doctors and patients will know that, up to a certain point at least, half the costs of health care have to be borne by the patient, rather than almost all simply being picked up by insurance or the government. Therefore, both doctors and patients will have an incentive to keep costs within reasonable bounds (doctors will because otherwise they may lose patients to those who do keep costs within reasonable bounds). It has been argued that the demand for medical services is so price inelastic that not even cost-sharing would make a significant difference in the fees charged. Perhaps so, in which case other means of keeping costs within reasonable bounds will have to be put into effect. Recent studies suggest, however, that cost-sharing does make a significant difference (Newhouse, Manning, Morris *et al.* 1981; Scitovsky and McCall 1977; Beck 1974; Reinhardt 1987).

Aside from cost-sharing, another feature of this programme will help prevent costs from escalating as in 1990. In the USA in 1990, a vast amount of all health-care costs were merely administrative—that is, costs associated, directly or indirectly, with determining by whom and how health-care costs are to be paid. These extensive administrative costs were largely a consequence of a reliance upon private insurance companies—1,500 of them—with their endless jumble of different policies, screening of would-be policy holders, army of costly insurance salespeople, claim adjusters and administrative personnel, advertising expenses, and bewildering array of paper work. Universal, comprehensive, income-related, cost-sharing greatly simplifies administration (same coverage for all, no need to screen would-be policy holders, simplified paper work, etc.), and it eliminates entirely the huge costs of competition (advertising, salespeople, etc.), all of which cut administrative costs tremendously. The arch-conservative dogma that the private sector can do anything at all more efficiently than government is, quite simply, false.

By the way, in contemplating how escalating medical costs can be controlled, consider this. Over a quarter of the 1990 US Medicare budget, or approximately twenty billion dollars, went to maintain patients in their last year of life, most of that in their last month (Halper 1991: 146). Many people believe that, if medical costs are ever to be controlled, it will be necessary to cut corners by rationing expensive medical treatment (e.g. Aaron and Schwartz 1990), which will mean that not everyone who can benefit from expensive treatment will be allowed to receive it. But would it not be infinitely more humane to 'cut corners' instead by permitting active euthanasia for the terminally ill who genuinely want it, rather than forcing them to endure these final few, costly months, no matter how excruciating their pain and suffering?

An alternative to a national *tax*-based—or 'single-payer'—health-care programme, such as advocated here, is a national *employer*-based programme, a programme requiring all employers to purchase, from private insurance companies, health insurance for their employees. Since relatively little, if any, tax money is required to finance an employer-based programme, with such a programme politicians need not risk incurring public disfavour by imposing a significant tax burden upon people. And, with an employer-based programme, politicians need not fear the wrath of the powerful insurance industry either. So an employer-based programme is the 'easy' way out for politicians, just like minimum wages are the 'easy' way out for politicians to provide relief from poverty.

Unfortunately, an employer-based programme, just like minimum wages, is not the best way. The costs to employers of such a programme are enormous. To the extent at least that these costs are not recoverable by employers from savings in taxes, they will either be passed on to employees in the form of lower wages than otherwise, or they will not. If they are, this will amount to a serious loss in pay falling largely upon the very people who can least afford it. If, on the other hand, these costs are not passed on to the employees, then they will have the effect of a significant increase in the minimum wage, which will, in turn, increase unemployment and consumer prices. And, to the extent that it increases consumer prices, it will mean more difficulty competing with foreign firms not burdened with these costs.

Moreover, an employer-based programme, relying as it does upon still more private insurance coverage, may do relatively little to curb the escalation of medical costs attributable to the 'cost-is-no-object-since-my-coverage-pays-all' mentality. And, clearly, it will not do as much as would a tax-based programme to curb those medical costs attributable to the enormous administrative costs associated with private insurance coverage (see above). So, although the tax burden from a tax-based programme is indeed significant, what many people fail to realize is that the insurance premiums necessary for financing an employer-based programme impose an even greater burden.

Finally, any regulations designed to achieve a certain minimum comprehensiveness with an employee-based programme create a two-tier system of health care: the rich supplement the minimum coverage with more adequate coverage, while the poor, along with their children, are left with second best. A comprehensive programme providing an equally high standard of coverage for all, as is called for by the ideal of equal opportunity, will almost surely never be realized except through a tax-based programme. And, in any case, building universality and even a minimum comprehensiveness into an employer-based programme will result in legislation that is far more complex than any that may be necessary for a tax-based programme, and let us never forget this basic fact of political life: unnecessary complexity, is, more often than not, no more than a camouflage for greedy, special interests.

But we need not pursue the details of alternative health-care programmes any further here. The general point is that, for protection against extraordinary expenses, a health-care programme is to be part of the overall antipoverty policy being set out here. And this programme is to be universal, comprehensive, and tax-based—not poverty-related, incomplete, or employer-based.

5.6.3 Education

Let us turn now to the third extraordinary expense for which some provision should be made: education. Industrialized countries already provide public education from kindergarten (or first grade) to high school that is available to rich and poor alike. In most of these countries, especially in the USA, much can and should be done to improve these public schools (e.g. see Thurow 1992: ch. 9), but here is not the place to pursue this matter. Since education from kindergarten through high school is already available to everyone, the major gaps for the poor are therefore after high school and before kindergarten. Moreover, adequate post high-school education and pre-kindergarten, or preschool, care are these days crucial: post high-school education because it opens up opportunities more than ever before, and preschool care not only because it can, potentially, afford children valuable stimulation and training difficult for them to receive otherwise, especially if they are living close to poverty, but also because it enables the parent in single-parent households, and both parents in dual-parent households, to work. Allowing the single parent to work is especially crucial, since single parents unable to work because of having to care for preschool children account for much poverty, especially in the USA. Adequate governmental funding of preschool child care, higher education for every person qualified, and trade schools for those not attending college or that need job retraining is expensive. But for any country that takes seriously the ideals of work rather than handouts and equal opportunity rather than special privileges—and I think every country does—adequate governmental funding is essential. The difficult question is therefore not whether these funds for education should be made available. They should. Rather, the difficult question is what governmental programme will work best.

Respect for the market suggests an income-related, cost-sharing programme, as in the case with medical expenses. What is proposed here, therefore, is this. Let a person get admitted to whatever college, university, or trade school he or she is able to get admitted to. Then the government will pay part of the costs and this person, or the person's parents, will pay the rest. The general idea is for the government's share to be such that it decreases gradually, and uniformly, as household income increases so as to make educational expenses more income neutral than at present, with the rate of this decrease being such that a family having an income that is about average will end up bearing approximately one-half of the costs and the government approximately one-half.

Since average US family income for 1990 was $36,520, what therefore would have been appropriate in the US for 1990 is a programme according to which families are to bear 1 per cent of the costs for every $500 of family income over the poverty line applicable to that family. Thus, in 1990, a family with average income and a rounded-off poverty line of $10,000 applicable to them would have had to have paid 53 per cent of the costs for the post high-school education of any children ([36,520 − 10,000]/500) and the government would have paid 47 per cent. (Note: in order to avoid an obvious loophole, a college-age child with rich parents is not to be permitted to gain a larger government subsidy than otherwise merely by becoming, for tax purposes, 'independent' with little or no income.) This programme is to be set up so that the amounts that families contribute *vis-à-vis* the government change automatically from year to year with major changes in the value of the dollar. And, to avoid any 'notch' problem, 1 per cent for every $500 could be computed instead as one-tenth of 1 per cent for every $5.

As an illustration, take two families of four, the poverty line applicable to them being $12,500. The first family's income is $17,500, and the second family's is $42,500. For a college costing $10,000 per year, the first family will pay $1,000 of this amount (10% of $10,000), and the second family, $6,000 (60% of $10,000), with the government paying, in each case, the remainder. And for a college costing $20,000 per year, the first family will pay $2,000, and the second family, $12,000. Thus, differences in costs and, presumably, in quality among different colleges and universities, are preserved, thereby allowing a person's choice of where to go depend upon both whether the person can gain admission and upon how much quality the person, or his or her family, chooses to pay for. In other words, the cost-sharing feature of this programme allows college, university and trade-school expenses to continue to be determined largely by the market, rather than having to be set entirely by the government.

Although sending children to college will still, even with this programme, necessitate a major financial sacrifice for most families, it will usually not be a devastating sacrifice. Moreover, by gradually increasing the percentage of the costs a family has to bear as its income increases, the extent of the sacrifice becomes more equal across all income levels. In other words, just as will medical care, post high-school education will indeed become more income neutral. And since, with this programme, the poor will no longer need as much in the way of scholarships or other private financial aid, more private aid can then go to the middle class.

Accordingly, this programme should make post high-school education at any institution for which one is qualified within the means of virtually every person.

Finally, government-backed student loans, although better than nothing, are not an adequate alternative. Wealth is opportunity, and debt is lack of opportunity. Student loans—if really sufficient for enabling the less well off to cover educational costs—will leave poorer young people with what may be a huge debt indeed, a burden richer classmates do not have to bear. For many years their savings may have to go merely to pay this debt off, while their richer classmates will be saving instead to finance their career dreams, retirement plans, or children's education. Moreover, such a burden may well discourage many deserving young people from ever pursuing higher education in the first place, especially those from ghetto homes where education may not be as highly valued as in middle- and upper-class homes. The income-related, cost-sharing programme, on the other hand, offers equal educational encouragement for all, and starts all off equally after graduation. This programme is therefore justified not only by appeal to the positive externalities of education, but by its greater realization than student loans of equal opportunity. In short, student loans serve best to supplement, not replace, income-related, cost-sharing.

Turning now to preschool child care, the best programme is, once again, an income-related, cost-sharing programme similar to the one for college education. Preschool child care is less costly than a college education. Yet families faced with preschool costs are generally younger than those faced with college costs and thus have less financial resources at their disposal. These factors—lower costs, yet younger families—tend to offset one another; so cost-sharing rates appropriate for college education tend to be appropriate for preschool care as well. Thus (just as with college expenses) it might have been appropriate, in 1990, for families to have taken on 1 per cent of their child-care costs for every $500 they made above the poverty line applicable to them, with the government contributing the rest. So as to avoid, once again, the notch problem, this could have been computed as one-tenth of 1 per cent for every $5. But, of course, these specific numbers are not as important as the general idea of government paying enough to make preschool care available to even the poorest of working parents, and parents paying enough to keep the costs of this care reasonably competitive and determined largely by supply and demand.

There should be more, however, to a governmental preschool programme than merely having the government share in the costs. Preschool care offers a unique opportunity for children to be stimulated and challenged in ways that can have lasting benefits. How successfully we are able at present to bring about lasting benefits through creative preschool training is debatable, but the evidence so far is encouraging (see e.g.

Osborn and Milbank 1987; Glazer 1986). Therefore, as part of the preschool programme, general government standards should be established for preschool curricula and for the training of preschool personnel, standards like those established for school curricula and for the training of public schoolteachers. These standards, incidentally, are not to prohibit anyone not meeting them—such as, say, a friend or relative—from accepting pay from a parent for preschool care; they are only to prohibit the parent from recovering part of this pay from the government. Since poor families often live in unfavourable conditions, children from poor families often do not have as much favourable early childhood stimulation and training as do children from more well-to-do families. It is to be hoped that, with the proper government standards, preschool care can be much more than just baby-sitting, that it can do much to close the gap in early childhood stimulation and training between rich and poor, thereby creating greater equality of opportunity.

And more than even these benefits is needed. An adequate, national maternity leave policy, and something that addresses the problem of so-called latch-key children (i.e. children who return from school before their parents come home from work) are needed as well, and are to be part of the antipoverty policy being set out here. But any details about these matters are beyond the scope of this enquiry.

In sum, the educational benefits proposed here (preschool, college, and occupational) not only constitute a redistribution of wealth, but they all contribute to capital accumulation—*human* capital and *knowledge* capital—which, in turn, contributes to economic growth. And these benefits prepare people for jobs. Thus, with these educational benefits, we have one place where all three components of the antipoverty policy being set out here—the redistributive, the jobs, and the growth components—coincide.

5.7 JOBS

The focus, so far, has been on the redistributive component of the policy being set out here. It is time now to turn specifically to the jobs component; that is, to measures for creating job opportunities. As we have seen, the policy being set out here provides no redistributive benefits for able-bodied people who do not work because they cannot find a job. That is, it provides no benefits other than those from unemployment insurance and the medical and educational programmes described above (including job-training programmes). This is far from full, comprehensive support. Unfortunately, those who do not work because they cannot find a job cannot, by any known means, be distinguished by the government from those who do not work simply because they do not want to work. Therefore, to provide full, comprehensive support for those who do not work because they cannot find a job would entail full, comprehensive support, as well, for those who do not work simply because they do not want to work. And full, comprehensive support for those who simply do not want to work is something that no society can afford. But if there is to be less than full support for those who cannot find a job, it is crucial that there be very few such people. It is, in other words, crucial that, for virtually anyone who wants a job, a job is available.

Of course if the programmes already described do, in fact, succeed in keeping the working population above the poverty line, then this by itself should help. Poor people are a drag upon the economy; if instead they are above the poverty line, they have more purchasing power, which in turn supports more jobs. But generalizations such as this are not enough. Adequate job opportunities require more specific measures. And we must keep in mind that creating job opportunities is a two-way process. It is not just a process of making jobs available for people, but also of making people available for jobs. What I mean by making people available for jobs is removing barriers preventing people from filling jobs that are already available, barriers such as lack of training, child-care responsibilities, and stifling inequalities of opportunity. Measures for both making more jobs available, and for removing barriers to filling available jobs are part of the jobs component of the policy being set out here. These measures include the following.

First, minimum wages are to be abolished. How many job opportunities this will create is debatable, but most agree that it will create a not insignificant number. Moreover, the refundable earned income credit proposed here makes it reasonably worthwhile, financially, for people to take these low-paying jobs.

Secondly, the income-related, cost-sharing preschool programme already described (Sect. 5.6) is to be designed so as to generate a vast system of regulated preschool child-care centres—a system large enough to assure affordable preschool care for the children of every parent that desires it, a system comparable in scope to the public school system itself. This preschool programme, as we saw, is to be combined with adequate maternity leaves and programmes for latch-key children. The beneficial effect of these programmes upon employment is threefold. First, and most obviously, it lifts the burden of child care for single parents, enabling them to work, and allows both parents in two-parent households to work. Secondly, the staffing needs of a vast system of preschool centres comparable to the public school system will create numerous new jobs. Third, and least obvious, the stimulation and training that children may receive in these centres, especially children who would not have received much early stimulation and training otherwise, may have long-term benefits that will make these children more employable as adults. These first two beneficial effects are certain, the last is more speculative, yet is potentially the most beneficial effect of all.

In general, the entire antipoverty programme set out here is designed to create more equality of opportunity and self-reliance. These goals are to be achieved not only by making highly stimulating preschool centres readily available to poorer families, but also by making higher education and vocational training more readily available, by assuring everyone adequate medical care, and by assuring working families sufficient income for basic necessities. It is reasonable to conclude that the combined effect of all these programmes will indeed be more equality of opportunity and self-reliance—or, in other words, a more thorough development of the earning potential of the poor—all of which should eventually pay off in a more healthy, faster growing economy, with fuller employment and more equal income. Moreover, although I shall not be discussing it here, this antipoverty policy is to include a growth component, which, if successful, should contribute to the creation of jobs as well.

But if these various measures did not increase employment opportunities sufficiently, another measure remains available, one that involves a small, structural change in current free-enterprise systems, but has perhaps the greatest job creation potential of all. I am referring to transforming current systems into 'share' economies, as proposed by Martin Weitzman (1984; for a more technical formulation of the idea, see Weitzman 1983). A share economy is simply an economy where industrial workers do not receive all of their pay in the form of fixed per hour, per week, or per month wages, but receive part of it instead in the form of a predetermined share in the firm's overall revenue, this share being a type of bonus.

Let me explain briefly how a share economy featuring income in the form of bonuses might be connected with the creation of jobs. Generally, the more workers a firm hires, the lower will be the marginal (i.e. the additional) revenue that the firm gets from each new worker. Yet the marginal costs of each new worker at any given skill level—these costs being the worker's salary—remain the same (a firm cannot pay each additional worker a little less so as to compensate for diminishing marginal productivity; rather, salaries must remain constant). In principle, it is to a firm's advantage to keep hiring additional workers up to the point of diminishing marginal returns; that is, up to the point where the cost to the firm of hiring an additional worker would exceed the revenue from that worker. Any new worker hired after this point would cost more than he or she was worth. Given that a firm's top priority is to maximize overall revenue, a firm would be willing to hire more workers once this point had been reached only if something could be done either to increase the marginal revenue from new workers or decrease their marginal costs. The general idea of a share economy is to bring about the latter—that is, a decrease in the marginal costs of new workers—thereby 'raising' a firm's point of marginal diminishing returns so that it can then hire more workers profitably.

The share economy decreases the marginal costs of new workers in the following way. As already pointed out, in a share economy part of what workers get paid takes the form of a bonus that comes from dividing a certain, predetermined percentage of the firm's revenue among the workers. The percentage of the firm's revenue that is to be thus divided among the workers will have been determined by ordinary bargaining procedures. Let us assume, for purposes of illustration, that, according to the collective-bargaining agreement of some firm, the percentage of the firm's revenue that the workers are to have divided up among themselves as a bonus is such that it will account for approximately one-third of what they get paid, the other two-thirds taking the form of ordinary, fixed wages. The costs to this firm of hiring new workers will then be reduced by approximately one-third over what it would be if a worker's entire pay took the form of ordinary, fixed wages, because the approximately one-third of any new worker's pay that is to take the form of a bonus will cost the firm nothing at all. Since the percentage of the firm's revenue to be used for worker bonuses is fixed, a new worker's bonus will be derived entirely from slightly decreasing the bonus of each of the old workers.

Being as the cost of each new worker will be one-third less, the firm can then hire more new workers before reaching its point of diminishing marginal returns—that is, before reaching the point where the cost to the firm of a new worker exceeds the revenue from that worker. Finally, if many large firms, through similar 'share' agreements, have thus reduced the costs of hiring new workers, then many more new workers will be hired; in other words, many new jobs will have been created.

Notice, incidentally, that the 'share' idea could work only with traditional, not worker-controlled firms. Notice also, however, that if a share economy actually did work according to theory, it could conceivably bring about some of the same benefits as worker control while, at the same time, avoiding the potential worker-control problems of insufficient investment and 'political' discrimination (Sect. 4.6). Indeed, a share economy could conceivably end up being traditional capitalism's answer to worker control, at least in part. Perhaps the main benefit of worker control is that, with such a structure, management's main incentive will be to please workers, which should result in a dramatic improvement in how ordinary workers are treated. If, however, a share economy did succeed in making it profitable for firms to keep hiring as many workers as possible, then, as Weitzman suggests (1984), to hire and hold these workers, 'share' firms might be forced to compete just as fiercely with one another in treating workers well as firms do today in treating customers well, which could improve how workers are treated to an extent that would rival worker control. Moreover, a share economy could, in principle, match, or even exceed, worker control's emphasis upon across-the-board reductions in pay rather than destructive lay-offs in times of recession. On the other hand, a share economy probably could not rival worker control in reducing the huge income gap between management and ordinary workers, in spreading investment income throughout society more evenly and, generally, in creating a strong sense of worker identification with the form that could prove to be both highly motivating and fulfilling for the ordinary worker (Sect. 4.6).

But how, it might be asked, can a share economy be brought about short of forcing it upon workers? The way to bring it about, gradually and voluntarily, is through tax incentives: simply by passing legislation that makes all or part of a worker's pay that comes from a fixed percentage share in his or her firm's revenue tax free. It is true that, with a share agreement, each existing worker's income will diminish slightly with each new worker that is hired (since with each new worker there will be more workers with whom the existing workers must share a fixed percentage of revenue). But this slight loss in income could be more than compensated for by favourable tax treatment, thus, it would appear, making it in the best interests of all that share agreements be made.

One job-creation programme that I have not mentioned so far is that of government itself becoming an employer by creating jobs directly—a programme of 'workfare', where the government guarantees work for everyone who cannot find it in the private sector. This guarantee of work would be made possible by means of governmentally created public-service jobs—genuinely useful jobs in such areas as sanitation, recreation, protection, and supplementary education, areas that, typically, are not well covered by the private sector.

It is true that these governmentally created jobs could in principle be made low paying and unattractive enough so as to preclude any mass exodus from the private sector to these jobs. But the administrative apparatus for this programme, the workers' salaries, and so on, would be a burden upon taxpayers. And how extensive to make this programme, and how high exactly to make workers' salaries, create difficult 'line-drawing' problems. If these line-drawing problems were not solved exactly right, such a programme could well, in subtle ways, undermine the workings of the market, and end up doing more harm than good.

A satisfactory solution to some of these line-drawing problems might, perhaps, be not to have these jobs administered by the government, but instead to have the government contract with private firms (after competitive bidding) to perform the services in question. These private firms would then be the ones to do the hiring, set the salaries, and supervise the work.

But we would still be left with what is probably the most serious line-drawing problem of all: that of drawing the line at when exactly these programmes should be cut back or were no longer needed at all. And, even if this line could be drawn accurately, there would remain the practical problem of implementing it, for once a governmental programme is in place with interests already vested, dislodging those vested interests is, typically, very difficult. Thus there is the danger of governmentally created jobs continuing long after they are no longer necessary, needlessly burdening taxpayers and perhaps contributing to a serious deficit problem.

Yet if all the other measures set out here for creating jobs were not, in combination, sufficient, then a moderate programme of governmentally guaranteed jobs would be worth considering. But only as a last resort; first, the other measures set out here should be given every opportunity to be sufficient by themselves.

5.8 COMPARED WITH 1990 POLICY

Aside from the growth component, which is beyond the scope of this enquiry, we now have before us the overall antipoverty policy being set out here, illustrated by how it would have applied to the United States in 1990. Let us now examine the advantages it has over actual, US antipoverty policy in 1990 (as set out in Sect. 5.2).

As we saw, actual 1990 antipoverty policy was inconsistent; equally deserving families were treated differently, their benefits depending arbitrarily upon where they happened to live or whether they were headed by a single parent or not. The policy set out here, however, is not inconsistent; irrelevant factors such as being headed by a single parent rather than two parents do not determine a family's benefits. In short this policy seals up the 'cracks' into which many 1990 households fell, and seals them up not only with respect to cash benefits, but—and this is crucial—with respect to medical and educational benefits as well.

With 1990 policy, the level at which benefits were set was too low; over 10 per cent of US citizens remained in poverty, many of these in working families. The policy set out here, however, not only assures that all working families with up to three children remain out of poverty provided only that their income at least equals the standard exemption, but this policy also provides protection for everyone against major extraordinary expenses.

Actual 1990 policy, as we have seen, provided people with perverse incentives; it, in some states, provided parents with an incentive to remain single, and it provided those without work with incentives to remain without work. No perverse incentives arise from the policy set out here. This policy is, in effect, neutral as between being married and being single, and its benefits are carefully structured so that a person is always significantly better off working than not. Indeed, one must be working to get any refundable tax-credit benefits at all. And the more one earns, the greater, up to a point, one's benefits will be. Although, to be sure, this policy may, occasionally, give rise to unusually high, effective marginal tax rates, these excessively high rates will be relatively rare and will always be temporary, lasting at most only so long as one's children are, say, attending college as dependents.

Policy as of 1990, due to minimum wages and to the absence of an adequate system of preschool care, did not provide adequate opportunities for people to work. The policy set out here does. It provides opportunities through the abolishment of minimum wages (thereby creating more jobs), through the creation of an adequate system of preschool care along with new educational opportunities, through greater equality of opportunity in general and, if necessary, even through the establishment of a 'share' economy and of governmentally guaranteed jobs.

Moreover, the minimum wages of 1990 policy were inefficient in another way: they often increased the wages of the wrong people—namely, low-income members of very affluent families. The policy set out here hits the poverty target more accurately.

The AFDC programme, the crux of 1990 policy, was extraordinarily complex, and administered by fifty separate state bureaucracies in combination with a federal bureaucracy, which made its administration extremely costly. In-kind programmes such as food stamps, each with their own costly bureaucracy, added still more to the administrative costs of 1990 policy. The refundable tax-credit programme—which, according to the policy being set out here, is to replace all these programmes—is much simpler and is to be administered by a single, already existing bureaucracy (the country's tax-collecting agency), all of which makes its administration far less costly.

Notice, in particular, how efficiently most benefits under this programme can be distributed. The tax-collecting agencies of most countries already require large employers to withhold anticipated income taxes from the wages of their employees. These countries can simply do the reverse for anticipated benefits under the refundable tax-credit programme, requiring large employers to add these benefits to their employees' wages, and then subtract any amounts so added from the withheld taxes they are to transfer to the government. In this way, refundable credit payments can be made at regular intervals rather than in one lump sum, and without the creation of any costly new bureaucracy.

Policy as of 1990 was paternalistic and demeaning to the poor, whereas the policy set out here is not. In the first place, the medical and educational benefits (including the preschool benefits) of the policy set out here are available, although at a diminishing rate, to almost all; therefore, like social-security benefits, these benefits will not carry with them the stigma of welfare. Nor does this policy embody the paternalism of in-kind benefits that are not necessary as a hedge against extraordinary expenses. And the benefits of the refundable tax-credit programme are administered in complete confidence through the country's tax-collecting agency, just as are income taxes, and therefore need not be any more of an embarrassment to anyone than are current tax returns.

In general, actual 1990 antipoverty policy did not succeed very well either in making the poor more self-reliant or in creating more equal opportunity. It discouraged self-reliance through structuring benefits so as to perversely discourage work, and through job-destroying minimum-wage legislation that took work away. It failed in creating equal opportunity by leaving the poor, and their children with inadequate medical care, inadequate preschool care, and inadequate educational opportunities, not to mention inadequate income, far below that necessary even for minimal subsistence. The policy set out here is designed to remedy all of these deficiencies, thereby making the poor more self-reliant and opportunities more equal.

5.9 SOME OBJECTIONS

We have seen so far that the antipoverty policy set out here has much to recommend it. Before coming to any conclusions, however, let us first look at some potential objections. Since this policy is composed of a number of different programmes, each with many controversial features, I cannot here go into all potential objections. I shall, however, touch briefly upon a few of the most obvious and potentially serious objections.

It is sometimes objected that a wage supplement—which is what the refundable earned income credit set out here amounts to—would cause employers to pay workers substantially less than they otherwise would pay them, since employers would reason that the government will, with the wage supplement, make up for any cutback in pay. In fact, however, competition, and the workings of supply and demand, will largely insure that workers are paid about what their productivity warrants, regardless of whether there are wage supplements from the government or not. An employer who attempted to pay his employees much less than what their productivity warranted would find them leaving so as to work for employers who did pay them what their productivity warranted. Only if an employer were able to corner the market on labour, which obviously is impossible, or if there were vast numbers of people unemployed, which the policy set out here makes unlikely, would below par pay be a serious problem.

A more legitimate objection is that a refundable credit as generous as the one set out here might serve as a disincentive to work (even if less of a disincentive than AFDC). It might, for example, encourage low-income heads of a household to work each year only up to approximately the point where they had earned enough for their income, along with their benefits from this credit, to put them above the poverty line—which might take, for example, a mere six months; thereafter, they might simply stay home and live off what they had earned combined with these benefits.

It is not possible to know, a priori, how serious the problem of people coasting once they had surpassed the poverty line would be, but I suspect that it would not be serious. Having reached, or slightly passed, the poverty line hardly leaves one living in luxury. One's desires for consumption usually far exceed what it is possible to satisfy merely with close to poverty-line income and, after one already had a job that enabled many of these desires to be satisfied (remember: one must have a job to get any of these benefits to begin with), it is unlikely that one would, say, give this job up halfway through the year, thus leaving the desires unsatisfied. After all, heads of households today whose yearly income, without these benefits, is twice the poverty line do not normally, after their income has, say, exceeded the poverty line about halfway through the year, simply stop working so as to enjoy more leisure. There is little reason why those with refundable-credit benefits would behave any differently.

Another objection to the refundable credit is that it would give rise to enforcement problems because of the incentive it provides low-income people for overstating their incomes. For example, so as to entitle them to a large credit two women might claim that they each earn $6,000 a year doing each other's housework. Or a single woman with children who is living with her parents might claim she earns $6,000 a year from her father for what amounts to no more than emptying the waste-paper basket at his place of business every now and then.

The remedy for these kinds of tricks is for the government to look beyond the alleged income to determine if the job really is a legitimate one or merely a sham for purposes of getting a refundable credit. Tax law in some countries already provides many with an incentive for claiming false income—namely, to avoid high gift and inheritance taxes by alleging that a gift or bequest was really income. And, of course, current income-tax law everywhere provides people with an incentive for *under*stating their incomes. Yet, in spite of these incentives to cheat, governments have always been able to uncover and punish enough false claims so as to make tax law workable. I see no reason why it would not be the same with a refundable-credit law. I see no reason, in other words, why enough false credit claims could not be uncovered and punished so as to make this law workable also, at least as workable as past and current tax laws.

But we come now to the most obvious, and potentially most devastating objection to the antipoverty policy set out here: its cost. Can it be implemented at a cost that is within reason? And in considering this question we must keep in mind that, in the United States at least, with the urgent need to somehow get current deficit spending under control, there may be relatively little leeway at present in what counts as 'within reason'.

On the other hand, in considering this question we must not get so bogged down in details that we lose sight of the overall picture, which is this. The policy set out here brings about more equal opportunity and self-reliance. A consequence of more equal opportunity and self-reliance will be a greater overall realization of human potential throughout society. And a greater overall realization of human potential means greater economic productivity. This greater economic productivity, in turn, may well mean the policy set out here will more than pay for itself—in the long run. (And, of course, even more long-run gains in productivity may result from the growth component of this policy, which has not been discussed in any detail here.)

But long-run gains in productivity do not solve the problem of the policy's immediate costs. And, for an antipoverty policy as different from the current policies of most countries as the one being set out here, the immediate costs are difficult to estimate. Perhaps we can gain some perspective on costs in the United States from the fact that to provide all those below the poverty line with grants sufficient for taking them out of poverty would cost only about 5 per cent of what the federal government currently spends (Thurow 1985: 61). But the overall policy set out here, since it also provides cash benefits as well as medical and educational benefits to many who are above the poverty line, will cost considerably more than that. (Remember: not only are some benefits to those above the poverty line desirable for their own sakes, but they are necessary in order to avoid the 'notch' problem.) On the other hand, this policy eliminates many expensive, current governmental programmes. Moreover, programmes that this policy substitutes for those to be eliminated have not just a redistributive component, but a cost-savings component as well. Take, in particular, the proposed health-care programme that not only distributes health care more fairly, but more efficiently as well. Taking all medical expenses, including any in the form of taxes, into account, even the upper and middle classes may well, on the average, get by more cheaply with this programme than before. And the proposed refundable tax credit is considerably less expensive to administer than the combination of programmes it is to replace. Yet, in spite of these major savings, the immediate costs of the antipoverty policy set out here remain substantial and the objection based upon these costs must be taken very seriously.

Earlier, merely for purposes of illustrating the refundable tax credit, I assumed a flat income tax rate of 33⅓ per cent. In order for the overall policy set out here to be economically feasible, however, a flat tax rate might in fact be required—although, of course, not necessarily one at 33⅓ per cent. A flat tax rate might be required so as to prevent marginal tax rates from going so high as to be a serious disincentive for work and investment. A flat tax rate might, in other words, allow the necessary revenue to be raised without anyone being taxed at an excessively high marginal rate.

And the capacity for raising large sums without excessively high marginal rates is by no means the only advantage of a flat tax rate. A flat rate, applicable to all incomes, including capital gains, is simple, and the advantages of simplicity in a tax code can hardly be overestimated. In the USA, the extreme complexity of current tax law serves largely to provide loopholes for the rich, fatten the pocketbooks of special-interest groups, and pre-empt the services of thousands of tax consultants who otherwise could be more usefully employed elsewhere. It is, in short, a national disgrace. Another advantage of a flat rate is that of foreclosing many of the strategies that the rich currently use for avoiding taxes, such as bracket shifting and tax arbitrage. Moreover, a flat rate eliminates the so-called problem of bracket creep, the problem of people's taxes rising merely because inflation has pushed them into higher tax brackets. A flat rate eliminates the marriage penalty, which is the increase in taxes people suffer from being pushed into a higher bracket as a result of having gotten married and combining their incomes. A flat rate eliminates the greater tax burden suffered by people whose incomes vary greatly from year to year (thereby pushing them into excessively high brackets in high-income years). A flat tax eliminates incentives to pursue certain kinds of activities rather than others merely because of differences in the rates at which they are taxed (a flat rate, in other words, eliminates certain 'distortions', and thus inefficiencies, in the market). Finally, a flat rate enables income to be taxed at its source (e.g. at the corporate level) rather than at its final destination (i.e. individuals). Taxing at the source not only has administrative advantages, but also makes it possible to eliminate the so-called double taxation of corporate income.

This last point about eliminating the double taxation of corporate income is important. According to current tax law almost everywhere, corporations are taxed on what they earn, then, after this income has been distributed to shareholders in the form of dividends, it is taxed again, at ordinary personal income rates. Not only does this double taxation seem unfair, but it is inefficient. It has been estimated that eliminating the double taxation to which corporate income is subjected would bring an efficiency gain of 2–3 per cent of gross national product (Hansson 1987). Moreover, double taxation gives rise to an illegitimate motive or corporate mergers and other investments, this motive being merely that of avoiding double taxation by 'ploughing' corporate income back into the corporation (whether the venture be particularly promising or not). But since a flat tax rate makes it possible to tax at the source of income without thereby benefiting those in higher brackets or hurting those in lower brackets, with a flat tax rate this double taxation is easily eliminated. The government need only tax all corporate income at the same flat rate that individual income is taxed (i.e. it need only tax at the 'source'), and then allow all individual income in the form of corporate dividends to remain tax free.

The standard objection against a flat tax rate is, of course, that it increases the relative tax burden upon the lower-middle class, a group that can ill afford any such increase. But if a flat rate is established not by itself, but in conjunction with all the components of the policy set out here, then this objection loses its sting. First, the $6,000 exemption to which, with this policy, everyone is to be entitled turns a flat rate into one that is somewhat progressive anyway. If, for example, we assume a flat rate of 33⅓ per cent, then, because of this exemption, a person earning $9,000 will end up being taxed at a rate of only 11 per cent of his total income (i.e. 33⅓ per cent of $3,000), while a person earning a million dollars will end up being taxed at virtually the full 33⅓ per cent. Secondly, any flat tax rate, according to the policy set out here, is (in the USA) to be accompanied by abolishing FICA taxes and thus financing Social Security through ordinary taxes. Since FICA taxes, being regressive no less, are therefore even more harmful to lower-middle class taxpayers than a flat rate, abolishing FICA taxes will tend to offset any increase in the burden from a flat rate. Thirdly, and most important, those in the lower-middle class, along with the poor, are the main beneficiaries of the overall policy set out here. They, along with the poor, will enjoy a very significant refundable tax credit (thereby decreasing their tax burden) and will enjoy major medical and educational benefits—all of which should more than compensate for any increased burden from a flat rate. So, all things considered, it makes sense to include a flat tax rate in the overall package being proposed here.

But let us not deceive ourselves. With or without a flat rate, a genuinely effective antipoverty policy—at first, if not in the long run—will be costly. It will be so costly that, if those of us in rich countries are really serious about eliminating poverty in our country, as indeed we should be, then, to raise the necessary revenue, we will have to make some very hard choices.

On the one hand, we could choose to raise this revenue simply by increasing taxes. Most capitalist economies today can probably tolerate even a major increase in taxes without serious negative effects. Taxes in some capitalist countries, such as Sweden, are already considerably higher than in other capitalist countries, such as the United States, and those economies with the higher rates do not seem to be suffering as a result; their manufacturing output per hour worked and rate of growth compare very favourably with those with the lower rates.

To be sure, raising taxes must be done with some caution; it may at times have unexpected, negative ramifications upon investment and growth. Raising taxes may, for example, cause a greater amount of investment money to flow to special-interest tax shelters at the expense of more productive investments; it may even cause a certain shift away from investment in general into consumption. But if so, the remedy need not be to not raise taxes. Alternatively, the offending tax shelters can be abolished, or a shift from taxing income to taxing only consumption can be implemented (see below).

On the other hand, we could choose to raise the necessary revenue not by raising taxes, but instead by repealing tax deductions that benefit primarily the upper and middle classes—deductions such as those in the United States for interest payments on home mortgages—or (even better) by repealing costly tax benefits in support of special interests (see Sect. 3.6). We could choose, in other words, to increase the tax base rather than tax rates. And, from the standpoint of economic efficiency, not to mention tax-code simplicity, increasing the tax base is the preferable choice. It is estimated that if, in 1985, the numerous special deductions, tax credits, loopholes, and special-interest provisions of US tax law were repealed, an additional $400 billion could have been raised (Pechman 1984: 13).

Finally, we could choose to raise the necessary revenue by scaling down existing, costly governmental benefits for the affluent. In the United States, these benefits range from shameless, special-interest subsidies for certain industries to social-security benefits for the affluent that go far beyond what is necessary. If, for example, affluent Americans were paid only as much in social-security benefits as they had contributed through payroll taxes, this alone would generate about $60 billion yearly (Kaus 1992: 143).

No doubt the best choice would be some combination of all three—higher rates, larger tax base, and reduced benefits for the affluent—with heavy emphasis upon the latter two. At the same time—for the sake of stimulating savings and investment, and thus economic growth—the switch from taxing income to taxing consumption instead should finally be made. Indeed, taxing consumption rather than income, along with an effective statute requiring the government to balance its budget, might serve to anchor the growth component of this overall antipoverty policy. (For more on these two measures, see Seidman 1990: chs. 3 and 4.)

In any case, the necessary revenue can, in principle, be raised without going much beyond closing loopholes, repealing special-interest legislation, and scaling down unnecessary middle- and upper-class benefits and tax deductions. Although such measures will, no doubt, unleash squeals of protest from the well-off, they will not (let us face it) cause any economic catastrophe.

In the end, the issue of costs comes down to this. Do not the benefits of eliminating most poverty justify any monetary cost short of that which would cause serious losses in work incentives, overall productivity, or freedom? The benefits of eliminating most poverty include greater access for all to necessities and greater equality of opportunity. They include children being able to look forward to more from life than just the false pleasures of drugs and alcohol. Do not the luxury items—the expensive wines and fancy clothes—that the well off might have to sacrifice to pay this monetary cost pale in comparison with benefits such as these? Surely they do.

But can these benefits in fact be realized without serious losses in work incentives, overall productivity, or freedom? With the right antipoverty policy, they can be—as I have tried to show throughout this chapter. Although these benefits cannot be realized without short-term, economic sacrifices from the well off, they can be realized without serious losses in work incentives, overall productivity or freedom.

Since implementing this policy does, however, require short-term, economic sacrifices from the well off, it may not be politically feasible—at present. But is such a policy nevertheless worth working towards for the future? Not if, as many libertarians believe, freedom is an 'absolute' value—one that must never be compromised at all—for this policy does compromise somewhat the freedom of the rich to ignore the suffering of the poor. And not if, as many socialists believe, equality of income or wealth is an absolute value, for this policy does not equalize income or wealth. But if, as I believe, neither freedom nor equality is absolute; if, instead, we should try to achieve the best possible compromise among a number of values—including especially those of equal access to necessities, equal opportunity, freedom, and productivity—then, I suggest, something like the antipoverty policy set out here may indeed be worth working towards for the future.

But for the most satisfactory realization of these four values—especially that of equal opportunity—still more than an adequate antipoverty policy is needed. What more is needed, I shall argue in the next and final chapter, is the abolishment of inheritance as we know it today.

6
CAPITALISM WITH EQUAL OPPORTUNITY

6.1 INTRODUCTION

Old ways die hard. A social practice may be taken for granted for centuries before humanity finally comes to realize it cannot be justified. Take, for example, slavery. Another example is the inheritance of political power. For many centuries, throughout most of the world, the suggestion that political power should be determined by democratic vote rather than heredity would have been met with scorn; today we realize just how unjustified determining political power by heredity really is.

Although we no longer believe in the inheritance of political power, most of us still believe in the inheritance of wealth, of *economic* power. But might not the inheritance of economic power be equally unjustified? This is the question to be examined here. Inheritance involves property rights; so another way of putting this question is: Should property rights incorporate the practice of inheritance as it exists today?

I shall not address the question of whether individuals may justifiably continue to take advantage of this practice as long as it exists, by continuing to bequeath and inherit property. (In my opinion, they may.) I address only the justifiability of the practice (or institution) itself. Finally, I shall, for the most part, be using the word 'inheritance' here to refer to any large amount one is given (as opposed to earns or wins), whether it be, technically, a bequest or a gift.

Many people support inheritance because they believe it to be essential to capitalism. In Section 6.2 of this chapter, I try to show that, far from being essential, inheritance is actually inconsistent with capitalism. Or, to be more exact, I try to show that it is inconsistent with fundamental values, or ideals, that underlie capitalism. For those, such as myself, who share these values, its inconsistency with them is prima-facie reason for abolishing inheritance. But prima-facie reasons for doing something can be overridden if the objections to doing it are strong enough. So, against the backdrop of a specific proposal for abolishing inheritance set out in Section 6.3, I examine, in Section 6.4, what I take to be the most important objections to abolishing inheritance. In Section 6.5, I propose a compromise that is, I think, preferable to the more stringent proposal for abolishing inheritance set out in Section 6.3. In Section 6.6, I set out the overall conclusions of this book. But let me begin, in this section, by presenting some useful background information about inheritance, and the distribution of income and wealth, in a typical capitalist country. Although this background information happens to come from the United States, the arguments and conclusions presented here are meant to apply anywhere.

Family income in the United States today is not distributed very evenly. The top fifth of American families receives approximately 44 per cent of all family income, while the bottom fifth receives approximately 4.4 per cent (*Background Material and Data on Programs within the Jurisdiction of the Committee on Ways and Means* 1989: 987; figures are adjusted for family size and do not count capital gains).

But, for obvious reasons, a family's financial well-being does not depend upon its income as much as it does upon its wealth, just as the strength of an army does not depend upon how many people joined it during the year as much as it does upon how many people are in it altogether. So if we really want to know how unevenly economic well-being is distributed in the United States today, we must look at the distribution not of income, but of wealth.

Although—quite surprisingly—the government does not regularly collect information on the distribution of wealth, it has occasionally done so. The results are startling. One to two per cent of American families own from around 20 to 30 per cent of the (net) family wealth in the United States; 5 to 10 per cent own from around 40 to 60 per cent (Avery *et al.* 1984: 865; Projector 1964: 285). The top fifth owns almost 80 per cent of the wealth, while the bottom fifth owns only 0.2 per cent (Thurow 1976: 33). So while the top fifth has, as we saw, about ten times the income of the bottom fifth, it has about 400 times the wealth. Whether deliberately or not, by regularly gathering monumental amounts of information on the distribution of income, but not on the distribution of wealth, the government succeeds in directing attention away from how enormously unequal the distribution of wealth is, and directing it instead upon the less unequal distribution of income. But two things are clear: wealth is distributed far more unequally than is income and this inequality in the distribution of wealth is enormous. These are the first two things to keep in mind throughout this discussion of inheritance.

The next thing to keep in mind is that, although estate and gift taxes in the United States are supposed to redistribute wealth and thereby lessen this inequality, they do not. Before 1981 estates were taxed, on an average, at a rate of only 0.2 per cent—0.8 per cent for estates over $500,000— hardly an amount sufficient to cause any significant redistribution of wealth (Thurow 1971: 127). And, incredibly, the Economic Recovery Act of 1981 *lowered* estate and gift taxes.

Of course the top rate at which estates and gifts are allegedly taxed is far greater than the 0.2 per cent rate, on the average, at which they are really taxed. Prior to 1981, the top rate was 70 per cent, which in 1981 was lowered to 50 per cent. Because of this relatively high top rate, the average person is led to believe that estate and gift taxes succeed in breaking up the huge financial empires of the very rich, thereby distributing wealth more evenly. What the average person fails to realize is that what the government takes with one hand, through high nominal rates, it gives back with the other hand, through loopholes in the law. Lester Thurow writes,

it is hard to understand why we go through the fiction of legislating high nominal rates and then nullifying them with generous loopholes—unless someone is to be fooled. The most obvious purpose of high nominal rates and low effective rates is to use the high nominal rates as a smokescreen to hide the transfer of wealth from generation to generation. (Thurow 1976: 100–1)

I do not know if the US government deliberately intends the law on estate and gift taxation to be deceptive but, due to the complications, exceptions, and qualifications built into this law, it is deceptive and, more seriously still, it is ineffective as a means of distributing wealth more evenly. Indeed, as George Cooper shows, estate and gift taxes can, with the help of a good attorney, be avoided so easily they amount to little more than 'voluntary' taxes (Cooper 1979). As such, it is not surprising that, contrary to popular opinion, these taxes do virtually nothing to reduce the vast inequality in the distribution of wealth that exists today.

Once we know that estate and gift taxes do virtually nothing to reduce this vast inequality, what I am about to say next should come as no surprise. This vast inequality in the distribution of wealth is (according to the best estimates) due at least as much to inheritance as to any other factor. Once again, because of the surprising lack of information about these matters, the extent to which this inequality is due to inheritance is not known exactly. One estimate, based upon a series of articles appearing in *Fortune* magazine, is that 50 per cent of the large fortunes in the United States were derived basically from inheritance (Smith 1957; Louis 1968, 1973). But by far the most careful and thorough study of this matter to date is that of John A. Brittain (1978). Brittain shows that the estimate based upon the *Fortune* articles actually is too low (1978: 14–16); that a more accurate estimate of the amount contributed by inheritance to the wealth of 'ultra-rich' males is 67 per cent (1978: 99). In any case, it is clear that, in the United States today, inheritance plays a large role indeed in perpetuating a vastly unequal distribution of wealth. This is the final thing to keep in mind throughout the discussion which follows.

6.2 INHERITANCE AND CAPITALISM

Capitalism (roughly speaking) is an economic system where (1) what to produce, and in what quantities, is determined essentially by supply and demand—that is, by people's 'dollar votes'—rather than by central planning, and (2) capital goods are, for the most part, privately owned. In the minds of many today, capitalism goes hand in hand with the practice of inheritance; capitalism without inheritance, they would say, is absurd. But, if I am right, the exact opposite is closer to the truth. Since, as I shall try to show in this section, the practice of inheritance is incompatible with basic values or ideals that underlie capitalism, what is absurd, if anything, is capitalism *with* inheritance.

Before proceeding, however, let me say a few brief words about ideals, or values, in general. Ideals, or values (and I use these terms interchangeably here) serve to delineate what is good; what, consequently, is to be striven for, and to be striven for even though in no way capable of ever being achieved fully. Take, for example, the ideal of the medical profession that everyone be in a state of perfect health. Ideals can be either absolute or prima facie. Absolute ideals admit of no compromise; they are to be realized as fully as possible no matter what the cost. Prima-facie ideals do admit of compromise; they are to be realized only to the extent that they do not conflict with, for example, more fundamental ideals. None of the ideals to be discussed here are absolute; all are merely prima facie. Finally, political values or ideals such as the ones we shall be considering, are, in my view, related to political morality, roughly, as follows. The most fundamental standard for evaluating governmental behaviour is that of the general welfare, the pursuit of which is constrained only by rights that people have against the government and other norms of political morality. Values or ideals, delineating, as they do, what is to be striven for, serve as general guides to what is in the general welfare. Accordingly, any governmental policy that is contrary to legitimate ideals is, other things being equal, unjustified. But if the contravened ideals are only prima facie, then the policy contrary to them can be shown to be justified after all, by showing that other things are not equal; by showing, in other words, that the policy, although contrary to these ideals, is required for other, more weighty reasons. But if it cannot be shown that the policy is required for other, more weighty reasons, then we must conclude that the policy is indeed unjustified.

I do not try to show here that the ideals underlying capitalism are worthy of support; I only try to show that inheritance is contrary to these ideals. And if it is, then from this it follows that, *if* these ideals are worthy of support (as, incidentally, I think they are), then we have prima-facie reason for concluding that inheritance is unjustified. What then are these ideals? For an answer, we can do no better than turn to one of capitalism's most eloquent and uncompromising defenders: Milton Friedman.

6.2.1 Distribution According to Productivity

The point of any economic system is, of course, to produce goods and services. But, as Friedman tells us, society cannot very well compel people to be productive and, even if it could, out of respect for personal freedom, probably it should not do so. Therefore, he concludes, in order to get people to be productive, society needs instead to entice them to produce, and the most effective way of enticing people to produce is to distribute income and wealth according to productivity. Thus we arrive at the first ideal underlying capitalism: 'To each according to what he and the instruments he owns produces' (Friedman 1962: 161–2).

Obviously, inheritance contravenes this ideal. For certain purposes, this ideal would require further interpretation; we would need to know more about what was meant by 'productivity'. For our purposes, no further clarification is necessary. According to any reasonable interpretation of 'productivity', the wealth people get through inheritance has nothing to do with their productivity. And one need not be an adherent of this ideal of distribution to be moved by the apparent injustice of one person working eight hours a day all his life at a miserable job and accumulating nothing, while another person does little more all his life than enjoy his parents' wealth and inherits a fortune.

6.2.2 Equal Opportunity

For people to be productive, however, it is necessary not just that they be motivated to be productive, but that they have the opportunity to be productive. This brings us to the second ideal underlying capitalism: equal opportunity—that is, equal opportunity for all to pursue, successfully, the occupation of their choice. According to capitalist ethic, it is OK if, in the economic game, there are winners and losers, provided everyone has an 'equal start'. As Friedman puts it, the ideal of equality compatible with capitalism is not equality of outcome, which would discourage people from realizing their full productive potential, but equality of opportunity, which encourages people to do so (1979: 131–40).

Naturally this ideal, like the others we are considering, neither could, nor should, be realized fully; to do so would require, among other things, no less than abolishing the family and engaging in extensive genetic engineering (see e.g. Williams 1962). But the fact that this ideal cannot and should not be realized fully in no way detracts from its importance. Not only is equal opportunity itself an elementary requirement of justice, but, significantly, progress in realizing this ideal could bring with it progress in at least two other crucial areas as well: those of productivity and income distribution. First, the closer we come to equal opportunity for all, the more people there will be who, as a result of increased opportunity, will come to realize their productive potential. And, of course, the more people there are who come to realize their productive potential, the greater overall productivity will be. Secondly, the closer we come to equal opportunity for all, the more people there will be with an excellent opportunity to become something other than an ordinary worker, to become a professional or an entrepreneur of sorts. And the more people there are with an excellent opportunity to become something other than an ordinary worker, the more people there will be who in fact become something other than an ordinary worker or, in other words, the less people there will be available for doing ordinary work. As elementary economic theory tells us, with a decrease in the supply of something comes an increase in the demand for it, and with an increase in the demand for it comes an increase in the price paid for it. An increase in the price paid for it means, in this case, an increase in the income of the ordinary worker *vis-à-vis* that of the professional and the entrepreneur, which, surely, will be a step in the direction of income being distributed more justly.

And here I mean 'more justly' even according to the ideals of capitalism itself. As we have seen, the capitalist ideal of distributive justice is 'to each according to his or her productivity'. But, under capitalism, we can say a person's income from some occupation reflects his or her productivity only to the extent that there are no unnecessary limitations upon people's opportunity to pursue, successfully, this occupation—and by 'unnecessary limitations' I mean ones that either cannot or (because doing so would cause more harm than good) should not be removed. According to the law of supply and demand, the more limited the supply of people in some occupation, then (assuming a healthy demand to begin with) the higher will be the income of those pursuing the occupation. Now if the limited supply of people in some high-paying occupation is the consequence of a 'natural' scarcity—a scarcity that is not the result of unnecessary limitations upon opportunity, but is the result instead of few people having the unborn capacity to pursue this occupation, or of few people freely choosing to do so—then (it is fair to say) the high pay does reflect productivity. Willingness or capacity to do what few people are willing or have the capacity to do is socially valuable, and those who in fact do it are therefore making an unusually valuable contribution; they are, in other words, being highly productive. But if, on the other hand, the limited supply of people in some high-paying occupation is the result of unnecessary limitations upon people's opportunity to pursue that occupation, then the scarcity is an 'artificial' one, and the high pay can by no means be said to reflect productivity. It can instead be said to reflect exploitation (cf. Miller 1989: ch. 7). The remedy is to remove these unnecessary limitations; in other words, to increase equality of opportunity. To what extent the relative scarcity of professionals and entrepreneurs in capitalist countries today is due to natural scarcity, and to what extent to artificial scarcity, no one really knows. I strongly suspect, however, that a dramatic increase in equality of opportunity will reveal that the scarcity is far more artificial than most professionals and entrepreneurs today care to think.

If my suspicions are correct, a dramatic increase in equality of opportunity not only would be desirable for its own sake (since equal opportunity is itself an elementary requirement of justice) but would also be desirable for the sake of greater productivity, a more equal distribution of income, and fuller realization of the ideal of distribution according to productivity. Indeed, a dramatic increase in equality of opportunity would, I suspect, do more to meet the objections that many throughout the world today have against capitalism than could anything else.

That inheritance violates the (crucial) second ideal of capitalism, equal opportunity, is, once again, obvious. Wealth is opportunity, and inheritance distributes it very unevenly indeed. Wealth is opportunity for realizing one's potential, for a career, for success, for income. There are few, if any, desirable occupations that great wealth does not, in one way or another, increase—sometimes dramatically—one's chances of being able to pursue, and to pursue successfully. And to the extent that one's success is to be measured in terms of one's income, nothing else, neither intelligence, nor education, nor skills, provides a more secure opportunity for 'success' than does wealth. Say one inherits a million dollars. All one then need do is purchase long-term bonds yielding a guaranteed interest of 10 per cent and (presto!) one has a yearly income of $100,000, an income far greater than anyone who, at present, toils eight hours a day in a factory will probably ever have. If working in the factory pays, relatively, so little, then why, it might be asked, do not all these workers become big-time investors themselves? The answer is that they are, their entire lives, barred from doing so by a lack of initial capital which others, through inheritance, are simply handed. With inheritance, the old adage is only too true: 'The rich get richer, and the poor get poorer.' Without inheritance, the vast fortunes in capitalist countries today, these enormous concentrations of economic power, will be broken up, allowing wealth, and therefore opportunity, to become distributed far more evenly.

6.2.3 Freedom

But so far I have not mentioned what many, including no doubt Friedman himself, consider to be the most important ideal underlying capitalism: that of liberty or, in other words, freedom. This ideal, however, is interpreted in different ways by different people. Earlier (Sect. 2.7 above) I distinguished between freedom in the 'narrow', and freedom in the 'broad' sense. Freedom in the narrow sense is simply the absence of coercion from others, including the government. Freedom in the broad sense is the absence not only of coercion, but also of any other kinds of constraints upon what we may desire to do, have, or be. Following Joel Feinberg (1980), we may classify these constraints as external positive ones (e.g. coercion from the government), external negative ones (e.g. a lack of money), internal positive ones (e.g. an addiction to drugs), and internal negative ones (e.g. a lack of information). And within each of these four, broad categories are many different kinds of constraints. One can thus be free in the broad sense in many different ways. That is, freedom in the broad sense has many different 'dimensions', with each dimension of freedom being the absence of a certain kind of constraint (see Sect. 2.7 above). Let us see how consistent inheritance is with freedom in each of these senses, starting with the narrow sense.

I think it is clear that when Friedman, and most other conservative defenders of capitalism, speak about the importance of freedom, they have in mind something like the 'narrow' sense, the sense in which freedom is simply the absence of coercion. Freedom in this sense, however, is not relevant here. Both inheritance and the abolishment of inheritance require laws and their enforcement. Accordingly, each requires a certain amount of governmental coercion. I do not know which, as between the governmental coercion necessary for supporting inheritance and that necessary for abolishing it, is greater in amount; I suspect the difference is insignificant. In reply it might be argued that, although the difference might be insignificant, coercion in support of inheritance would nevertheless be more justified, since coercion in support of inheritance would be in support of our fundamental property rights. But such an argument would merely beg the question; whether our fundamental property rights should include the practice of inheritance is the very point at issue (see Sect. 6.4, item 1 below). We need not, therefore, give freedom in the narrow sense any further consideration here.

Consider next freedom in the broad sense. Contrary to those who espouse freedom in the narrow sense, those who espouse freedom in the broad sense are often opponents of capitalism, and they use the broad sense of freedom to try to show how capitalism, with the vast inequalities of wealth to which it gives rise, is actually inconsistent with freedom. But those who espouse freedom in the broad sense need not be opponents of capitalism, since, arguably, capitalism, or some modified version of it, provides even the less well off with more freedom in the broad sense than does any other system (see Sect. 3.5 above). Indeed, it is freedom in the broad sense that is espoused by most liberal defenders of capitalism, those who, although they believe in the free market, are more sympathetic to governmental aid for the poor than are conservatives. Moreover, for the reasons explained earlier (Sect. 2.6), freedom in the broad sense is, for purposes of evaluating social institutions, the more appropriate ideal of freedom. Therefore, I include freedom in the broad sense among the ideals that underlie capitalism even though this ideal is not supported by all defenders of capitalism, and is, in fact, supported by many of its foes.

Let us now see how consistent inheritance is with this sense of freedom. Without inheritance people are no longer free to leave their fortunes to whomever they want and, of course, those who otherwise would have received these fortunes are, without them, less free to do what they want also.

But to offset these losses in freedom are at least the following gains in freedom. First, as is well known, wealth, generally speaking, has diminishing marginal utility. It also has diminishing marginal desire satisfaction (for the difference between utility and desire satisfaction, see Sect. 1.7 above). Diminishing marginal desire satisfaction means that, with each additional increment of wealth one gets, generally speaking the desires this additional wealth goes to satisfy will be less urgent or important to one than the desires the previous increment of wealth went to satisfy. Now let us assume for the sake of argument that *overall* wealth throughout society is not going to decrease as a result of distributing it more equally. Given this assumption, we may conclude from the diminishing marginal desire satisfaction of wealth that, the more equally wealth is distributed, the more overall desire satisfaction there is likely to be throughout society (for details, Sect. 3.5 above). But if the more equally wealth is distributed, the more the overall desire satisfaction, this means obviously that, the more equally wealth is distributed, the less the overall *constraints* upon desire satisfaction. Finally, the less the overall constraints upon desire satisfaction, the more the overall freedom (since freedom in the broad sense is, once again, the absence of constraints). Therefore, the more equally wealth is distributed, the more overall freedom there will be. So assuming that, as a result of abolishing inheritance, overall wealth will not decrease, an assumption I try to defend in Sect. 6.4 below, it follows that abolishing inheritance will increase freedom simply by distributing wealth more equally.

Moreover, a more equal distribution of wealth will contribute to breaking down class barriers, thus promoting 'social' freedom. And, in so far as political power is a function of wealth, a more equal distribution enhances political freedom as well.

Finally, abolishing inheritance also increases freedom by increasing equal opportunity. Relative to those who inherit significant funds, those who do not inherit significant funds clearly start life with what amounts to a significant handicap or constraint. Abolishing inheritance, and thereby starting everyone at a more equal level, leaves those who otherwise would have suffered from this constraint more free to attain whatever positions and occupations they may desire.

I submit that these gains in freedom—that is, those attributable to distributing wealth more equally and to more equality of opportunity—more than offset the loss in freedom resulting from the inability to give one's fortune to whom one wants. Abolishing inheritance is analogous to abolishing discrimination against blacks in restaurants and other commercial establishments. By abolishing discrimination, the owners of these establishments lose the freedom to choose the skin colour of the people they do business with, but the gains in freedom for blacks are obviously greater and more significant than this loss. Likewise, by abolishing inheritance the gains in freedom for everyone else are greater and more significant than the loss in freedom for the rich.

In general, all conservatives are suspicious of concentrating great economic power in the hands of government because of the threat this poses to freedom. These suspicions are well founded. Some conservatives are suspicious of concentrating great economic power in the hands of corporations. On the whole, these suspicions are well founded also; with the proviso that, because of economies of scale, certain of these concentrations are a necessary evil. No conservative, as far as I know, is suspicious of concentrating great economic power—via private fortunes—in the hands of individuals. It is time for conservative thinking to become more consistent: the truth is that all concentrations of great economic power threaten freedom, whether emanating from government, corporations, or individuals. So to the list of ideals with which inheritance is inconsistent we must add freedom in the broad sense.

To recapitulate: three ideals that underlie capitalism are 'distribution according to productivity', 'equal opportunity', and 'freedom'. Freedom is subject to either a narrow or a broad interpretation, but only the broad interpretation is relevant here. I do not claim these are the only ideals that may be said to underlie capitalism; I do claim, however, that they are among the most important. Inheritance, I have argued, is inconsistent with each of these ideals. Since they are among the most important ideals that underlie capitalism, I conclude that inheritance not only is not essential to capitalism, but is inconsistent with it.

For those who, like myself, are inclined to support these ideals, the inconsistency of inheritance with them creates a strong presumption that inheritance is unjustified. But only a presumption, for none of these ideals are absolute. There may, therefore, be objections to the abolishment of inheritance that are strong enough to override its inconsistency with these ideals (or to show that my conclusion about its inconsistency with the ideal of freedom to be premature). So, before we can make any final judgement about the justifiability of inheritance, we must examine the most significant of these objections. Yet to properly examine these objections, we should have a definite proposal for abolishing inheritance before us. I shall set out such a proposal next.

6.3 A PROPOSAL FOR ABOLISHING INHERITANCE

In this section, I set out a proposal for abolishing inheritance that incorporates the main features that I think such a proposal should incorporate. Although I shall, in Section 6.5, end up supporting a compromise proposal instead, before examining this compromise proposal it will be useful to see first the extent to which the more stringent proposal set out in this section can be justified.

First, this proposal for abolishing inheritance includes the abolishment of all large gifts as well—gifts of the sort, that is, which might serve as alternatives to bequests. Obviously, if such gifts were not abolished as well, any law abolishing inheritance could be avoided all too easily.

Of course we would not want to abolish, along with these large gifts, such harmless gifts as ordinary birthday and Christmas presents. This, however, raises the problem of where to draw the line. I do not know the best solution. The amount that current US law allows a person to give each year tax free ($10,000) is too large a figure at which to draw the line for purposes of a law abolishing inheritance. We might experiment with drawing the line, in part at least, by means of the distinction between, on the one hand, consumer goods that can be expected to be, within ten years, either consumed or worth less than half their current value and, on the other hand, all other goods. We can be more lenient in allowing gifts of goods falling within the former category since, as they are consumed or quickly lose their value, they cannot, themselves, become part of a large, unearned fortune. The same can be said about gifts of services. But we need not pursue these technicalities further here. The general point is simply that, so as to avoid an obvious loophole, gifts (other than ordinary birthday presents, etc.) are to be abolished along with bequests.

Next, according to this proposal, a person's estate is to pass to the government, to be used for the general welfare. If, however, the government were to take over people's property upon their death then, obviously, after just a few generations the government would own virtually everything—which would certainly not be very compatible with capitalism. Since this proposal for abolishing inheritance is supposed to be compatible with capitalism, it must therefore include a requirement that the government sell on the open market, to the highest bidder, any real property, including any shares in a corporation, that it receives from anyone's estate, and that it do so within a certain period of time, within, say, one year from the decedent's death. This requirement is, however, to be subject to one qualification: any person specified by the decedent in his will shall be given a chance to buy any property specified by the decedent before it is put on the market (a qualification designed to alleviate slightly the family heirloom/business/farm problem discussed below). The price to be paid by this person shall be whatever the property is worth (as determined by government appraisers, subject to appeal) and any credit terms shall be rather lenient. And, to prevent the 'fragmentation' of farmland through governmental sale, any such sale of farmland could be limited, by law, only to its sale as a whole with, perhaps, a restriction in the deed prohibiting any resale within a certain number of years except as a whole.

Finally, the abolishment of inheritance proposed here is to be subject to three important exceptions. First, there shall be no limitations at all upon the amount a person can leave to his or her spouse. A marriage, it seems to me, should be viewed as a joint venture in which both members, whether or not one stays home tending to children while the other earns money, have an equally important role to play; and neither, therefore, should be deprived of enjoying fully any of the material rewards of this venture by having them taken away at the spouse's death. And unlimited inheritance between spouses eliminates one serious objection to abolishing inheritance: namely, that it is not right for a person suddenly to be deprived, not only of his or her spouse, but also of most of the wealth upon which he or she has come to depend—especially in those cases where the spouse has, for the sake of the marriage, given up, once and for all, any realistic prospects of a career.

The second exception to be built into this proposal is one for children who are orphaned, and any other people who have been genuinely dependent upon the decedent, such as any who are mentally incompetent, or too elderly to have any significant earning power of their own. A person shall be able to leave funds (perhaps in the form of a trust) sufficient to take care of such dependents. These funds are to be used only for the dependent's living expenses, which include any educational or institutional expenses no matter how much. They are not, of course, to be used to provide children with a 'nest egg' of the sort others are prohibited from leaving their children. And at a certain age, say twenty-one (if the child's formal education has been completed), or upon removal of whatever disability has caused dependency, the funds are to cease. This exception eliminates another objection to abolishing inheritance—the objection that it would leave orphaned children, and other dependents, without the support they should have.

The third and final exception to be built into this proposal is one for charitable organizations—ones created not for purposes of making a profit, but for charitable, religious, scientific, or educational purposes. And, in order to prevent these organizations from eventually controlling the economy, they must, generally, be under the same constraint as is the government with respect to any real property they are given, such as an operating factory: they must, generally, sell it on the open market within a year.

A limit of some sort could be placed upon the amount a person would be entitled to give to charitable organizations; I am inclined, however, to oppose any limit. Allowing unlimited contributions will, to be sure, weaken one of the advantages of abolishing inheritance, that of providing government with a major, new source of revenue, one that serves to lessen the burden of income taxes. For rather than allowing their estates to pass to the government, many people will probably choose to leave their estates to charitable organizations. But even if they do, the advantage of lessening our tax burden will not be lost altogether, for if vast amounts are given to charitable organizations, these organizations can be expected to fund much of the welfare, medical research, scholarships for the poor, aid to education, and so on, that must (or should) now be funded by the government, thereby lessening our tax burden not by increasing governmental revenues, but by decreasing governmental expenses. And charitable organizations will become, even more than they are today, a healthy counterbalance to the power of government in these areas.

Among the objections to abolishing inheritance is that it would serve to dry up charitable giving and, indeed, leave our charitable instincts, which are among our most noble, with no (monetary) ways of being expressed. But with the unlimited exception for charitable giving built into this proposal for abolishing inheritance, charitable giving, far from drying up, will actually increase, and increase significantly. Thus the effect of this proposal upon charitable giving, far from providing an argument against abolishing inheritance, actually provides still another argument in favour of it.

We now have before us a specific proposal for abolishing inheritance. Let us turn to the objections.

6.4 OBJECTIONS

1. I shall begin the survey of objections to abolishing inheritance with the one that is weakest: the objection that abolishing inheritance is a violation

of property rights. The trouble with this objection is, quite simply, that it begs the question. Property rights are not normally viewed as being unqualified, nor certainly should they be. On the contrary, property rights normally are, and certainly should be, viewed as having built into them a number of qualifications or exceptions: an exception for taxes, an exception for uses which pose a danger of injury to others, an exception for eminent domain, and so on. As pointed out at the very beginning, the purpose of our investigation is precisely to determine whether we should recognize still another exception to property rights—an exception in the form of abolishing inheritance—or whether, instead, property rights should incorporate the practice of inheritance. Obviously we cannot determine this simply by slamming our fists on the table and insisting that property rights *do* allow inheritance, that abolishing inheritance *is* a violation of property rights. The only way to determine whether property rights should incorporate the practice of inheritance is the way we are proceeding here, the hard way: by patiently examining, one by one, the various pros and cons of abolishing inheritance and then, in the light of this examination, making as accurate an overall assessment as we can.

And lest my opponent claim this to be merely an appeal to narrow utilitarian considerations and hence (in his view) illegitimate, let me hasten to point out that the 'pros and cons' of which I speak need not be limited merely to considerations of utility. Considerations of right (other than the very one in question), of obligation, and of value are welcome additions to the debate. The *indirect* utilitarian view presupposed here (see Ch. 1 and Sect. 2.2) admits all such considerations, and more. The only thing not welcome is a demand that we merely accept my opponent's view of property rights without question.

2. Let us turn next to the reason Milton Friedman supports inheritance (Friedman 1962: 163–4; 1979: 136) and does so in spite of the fact that, as we have seen, it is incompatible with the values he himself says underlie capitalism. And, incidentally, Friedman is not alone in finding this particular objection to abolishing inheritance convincing; others who include Robert Nozick and F. A. Hayek (Nozick 1974: 237–8; Hayek 1960: 90–1; cf. Rawls 1971: 278). The argument upon which this objection is based proceeds somewhat as follows. Inheritance of property ('material' inheritance) is not the only source of unequal opportunity. Some people, for example, gain an unearned advantage over the rest of us by inheriting from their parents a beautiful singing voice, or keen intelligence, or striking good looks ('biological' inheritance). If we allow people to enjoy unearned advantages from biological inheritance, so the argument goes, it is only fair that we allow people to enjoy unearned advantages from material inheritance as well. We are told, in effect, that if we continue to allow one kind of unearned advantage to exist, we are, in all fairness,

committed to allowing all other kinds of unearned advantages to continue to exist also.

The fallacy in this way of arguing should be apparent. One might just as well insist that racial discrimination is justified by arguing that because we allow unearned advantages resulting from biological inheritance to continue to exist, we are, in all fairness, therefore committed to allowing unearned advantages resulting from racial discrimination to continue to exist also. To be sure, we do 'allow' unearned advantages resulting from biological inheritance to continue to exist because, first of all, we cannot eliminate them and, secondly, even if we could, we would not want to since the costs of doing so would outweigh the benefits. Unearned advantages resulting from material inheritance, on the other hand, can be eliminated. Perhaps the costs of doing so outweigh the benefits here as well. But, once again, we can determine whether this is so only the hard way: by a careful and patient investigation into what the pros and cons actually are. We must not be dissuaded from this task by the above quick, but fallacious argument with which Friedman and others tempt us.

3. Let us turn now to some objections of a more practical nature. The first of these is that, regardless of whether abolishing inheritance is justified or not, the simple truth is that the vast majority of people are solidly against doing so; therefore our discussion is a waste of time. The reply, of course, is that, unless popular opinions are sometimes challenged, how, after all, can we ever make any progress? What if, for example, no one in the old South had been willing to consider whether abolishing slavery was justified? If abolishing inheritance really is justified, chances are that people will eventually *come* to favour it; but the first order of business, clearly, is to decide if it really is justified.

But we must not dismiss this practical objection too quickly; this objection is related to still another practical objection, one that does deserve to be taken seriously: the objection that a ban on inheritance could never be adequately enforced. Some would argue that, by secret Swiss bank accounts, bogus exchanges, fake salaries, and simply by passing money under the table, large gifts could be made at any time in spite of a law abolishing them. Therefore, it would be concluded, no matter how justified the abolishment of inheritance might be, any law to that effect would be a futile gesture; as the sophomore is fond of saying, it might be good in theory, but would never work in practice.

Yet would a ban on inheritance actually be unenforceable? The possibility of gain through illegal gifts exists with current estate and gift tax law. Nevertheless, governmental investigations do succeed in uncovering such gifts often enough for successful enforcement. The reports of courts that handle tax litigation are filled with such cases. This suggests the government could, similarly, enforce a law abolishing inheritance.

But, it may be replied, current estate and gift tax law provides the wealthy with so many legal means of avoidance that (provided only that they are willing to hire a clever attorney) illegal means are not even necessary. A law abolishing inheritance, one not so easily avoided through legal means, would be an entirely different matter. With it, the 'need' to resort to illegal means would be far greater and, therefore, so would the difficulties of enforcement.

There is, no doubt, some truth in this reply. But so far we have overlooked an important factor: enforcement through informal social pressure, as opposed to formal governmental pressure. And by 'informal' social pressure, I mean more than just the pressure that potential social disapproval and ostracism exert upon us; I mean also the pressure our early training and social conditioning exert upon us, a pressure so great that most of us would not violate most laws even without any governmental enforcement of them. I doubt if governmental enforcement would be adequate enforcement for many laws without social pressure to go along with it. Take, for example, the legal prohibition upon alcoholic beverages in the United States which, in spite of intensive governmental enforcement, failed, and did so primarily because of a lack of popular support or, in other words, a lack of informal social pressure for compliance with the law. On the other hand, with the help of informal social pressure, I doubt if there are many laws the government could not enforce adequately, even if the government's role amounted to mere tokenism. Take, for example, current income-tax law; in spite of possibilities for cheating that rival those of a law abolishing inheritance, this law does, for the most part, work; and it works because, although few agree with the details of the law, most believe that, in principle, an income tax is justified. And if most people believed a law abolishing inheritance was justified then, through a combination of government pressure and informal social pressure, this law likewise would work well enough.

This, however, is where the fact that most people currently do not think such a law would be justified becomes relevant. Since most people currently do not think such a law would be justified, that such a law could be adequately enforced at the present time is indeed doubtful. And if it could not be adequately enforced, I must conclude that, currently, no such law should be passed. Except where absolutely necessary in order to protect minority rights, the imposition of extremely unpopular laws upon an unwilling public has no place in a democracy.

None of this, however, must be allowed to obscure an equally important point: the strength of a democracy depends upon there being the freedom and willingness to question the status quo, to constantly seek a better way. Although currently a law abolishing inheritance appears to have little public support, if such a law really is justified except for currently having little public support, I believe the necessary public support will eventually come—provided we retain the freedom and willingness to question the status quo.

4. We turn next to what is, I suppose, the most common objection to abolishing inheritance: the objection that, if people were not allowed to leave their wealth to their children, they would lose their incentive to continue working hard, and national productivity would therefore fall. In spite of the popularity of this objection, all the available evidence seems to indicate the contrary. For example, people who do not intend to have children, and therefore are obviously not motivated by the desire to leave their children a fortune, do not seem to work any less hard than anyone else. And evidence of a more technical nature leads to the same conclusion (McClelland 1961: 234–5; Fiekowsky 1959: 370–1): people, typically, do not need to be motivated by a desire to leave their children (or someone else) great wealth in order to be motivated to work hard.

Common sense tells us the same thing. The prospect of being able to leave one's fortune to one's children is, no doubt, for some people one factor motivating them to be productive. But even for these people, this is only one factor; there are usually other factors motivating them as well, and motivating them to such an extent that, even if inheritance were abolished, their productivity would be unaffected. Take, for example, professional athletes. If inheritance were abolished, would they try any less hard to win? I doubt it. For one thing, abolishing inheritance would not, in any way, affect the amount of money they would be able to earn for use during their lives. So they would still have the prospect of a large income to motivate them. But there is something else that motivates them to do their best which is, I think, even more important, and is not dependent on money: the desire to win or, in other words, to achieve that which entitles them to the respect of their colleagues, the general public, and themselves. Because of the desire to win, amateur athletes compete just as fiercely as professionals. Abolishing inheritance would in no way affect this reason for professional athletes doing their best either; they would still have the prospect of winning to motivate them. Businessmen, doctors, lawyers, engineers, artists, researchers—in general, those who contribute most to society—are not, with respect to what in the most general sense motivates them, really very different from professional athletes. Without inheritance, these people would still be motivated by the prospect of a sizeable income for themselves and, probably even more so, by the prospect of 'winning'; that is, by the prospect of achieving, or continuing to achieve, that which entitles them to the respect of their colleagues, the general public, and themselves.

But even if abolishing inheritance did lessen incentive by leaving people with no motivation to accumulate for their children, it would, in another respect, increase incentive. This can be illustrated by the often used race analogy. Consider two different hundred-metre races between Jones and Smith where, in race one, Jones is given a fifty-metre head start while, in race two, they start even. In which of these races will each be most likely to run his fastest? The answer, of course, is race two. In race one, Jones will figure that even if he does not run his fastest, he will win, while Smith will figure that even if he does run his fastest, he will lose; so very likely neither will run his fastest. What is true in this example is probably true in the 'game of life' as well: the more equal people's starting-points, the more incentive they will have to try hard. Since abolishing inheritance would do much to equalize people's starting-points, it should, in this way, increase people's incentives. And this increase, attributable to more equality of opportunity, would, I should think, more than make up for any decrease in incentive attributable to having no one to leave one's fortune to—if, indeed, there were any such decrease.

5. We come now to the most technical and, potentially, the most serious objection to abolishing inheritance: the objection that this would cause a substantial decrease in savings and investment, thus causing a serious reduction in capital which, in turn, would erode our standard of living (Hoover 1927; Johnson 1914; see also Tullock 1971). Abolishing inheritance, it is said, would reduce savings for two reasons. First, by breaking up large fortunes, it would reduce the number of people whose wealth far exceeded their capacity to consume, and who, therefore, were able to sink vast amounts into savings and investment. Without these vast amounts going into savings and investment, it is argued, overall savings and investment, and thus capital, would go down. Secondly, it is said that, not only would abolishing inheritance reduce people's *capacity* to save by breaking up large fortunes, it would also reduce people's *incentive* to save. Although, as we have seen, abolishing inheritance probably would not significantly affect people's incentive to produce, their incentive to save might well be affected. If people could not leave their wealth to their children, then rather than leave it to charity, or to the government, they might well decide to consume it instead. In short: abolishing inheritance would shift people's consumption–savings pattern more in the direction of consumption.

These points, although probably well taken, should not be exaggerated. Abolishing inheritance would distribute wealth more evenly, the relatively few enormous fortunes of today being replaced, in part, by a larger number of moderate fortunes. Thus any slack in investment attributable to a decrease in the *size* of the fortunes of those with enough to invest substantial amounts would, to some extent, be taken up by an increase in the *number* of people with fortunes large enough for them to invest substantial amounts. And, with the abolishment of inheritance, people's motives to save and invest would certainly not evaporate altogether

(Barlow *et al.* 1966: 31–3). People would, of course, still want to hold something back for a 'rainy day'. People would still want to save for their retirement (and no one knows for sure how many years one will need to be covered). Investments would still remain attractive aside from the savings motive; they combine the excitement of a gamble with the satisfaction of doing what is socially useful. And many people, especially the more wealthy, would still want to save for charitable purposes; to have a scholarship or perhaps a university building named after them, to support medical research, or even to establish a charitable foundation to carry on with some project in which they deeply believe. Another point to keep in mind is that most corporate investment—and corporate investment constitutes a large percentage of total investment—is generated by corporate income. Given the separation between ownership and management in large corporations today, it is unlikely that management would be influenced by the abolishment of inheritance to reduce the percentage of corporate income used for the replacement of capital and new investment. Indeed, if private funding became increasingly scarce, funding from corporate income might well take up some of the slack. In short, no one really knows for sure exactly how abolishing inheritance would affect savings and investment, and the effect might turn out to be far less than some critics think.

But the most important reply to the investment objection is this: even if abolishing inheritance significantly reduced investment, countermeasures to increase investment—and to do so at relatively little cost—are readily available. These countermeasures can be classified as either direct or indirect. Indirect methods include, for example, reducing the availability of consumer credit, requiring full funding of all pension plans, and taxing consumption. Direct methods take the form of a direct government subsidy of investment. But, it should be emphasized, not even direct methods need be at the cost of any governmental management of investments, which would be the first step towards socialism. Indeed, they need not even be at the cost of any governmental selection of investments, something which most supporters of the free market would find objectionable also. For example, the government might stimulate investment directly by partially guaranteeing loans to be used for creating new capital, the percentage of the loan to be guaranteed by law to be set at a figure no higher than necessary for stimulating the amount of investment desired throughout the economy. And by a requirement that the government not loan any money itself, just partially guarantee all and only those loans which entrepreneurs succeeded in procuring from independent investors and financiers, the government would be precluded from ever selecting which investments to stimulate. Another direct method of stimulating investment might take the form of governmental matching funds where, for any amount an entrepreneur is able to raise on his own, the government matches it with some additional sum which would be set by law at whatever would be necessary for stimulating the desired amount of investment.

At this point one might object that all direct methods, at least, have one crucial drawback: they would be very expensive. Perhaps so. (It would depend on how much of a need to support investment, if any, the abolishment of inheritance gave rise to, something which, once again, we have no way of knowing in advance.) We must remember, however, that, by abolishing inheritance, the government will experience either a major new source of revenue (people's estates) or a major reduction in its expenses (attributable to a major increase in private charity). Either way, the government is likely to end up with a substantial increase in funds that can be used to offset any such expenses.

But a discussion of the details and relative merits of different methods of stimulating savings and investment is beyond the scope of our present investigation. The point is simply that, as Lester Thurow says: 'If more savings are desired, they can be had' (1976: 107). And, as we have seen, they can probably be had at relatively little cost, whether the cost be measured in dollars, or in terms of governmental interference with freedom of investment. So, once again, we see that a major objection to abolishing inheritance turns out, upon closer examination, to be largely unsubstantiated.

Recall, incidentally, that I shall end up supporting a compromise proposal that does not entail the abolishment of inheritance altogether (Sect. 6.5). As we shall see, this compromise proposal provides all but a few of the very rich with virtually as much incentive to save as they have today. Thus, with this compromise proposal, the likelihood of substantiating any serious savings and investment objection is even less.

But it is now time to turn to some objections that can be substantiated. The only question is how serious these objections are.

6. First, there is the objection about administrative costs. As we saw, in order to prevent the government and charitable organizations from eventually owning virtually everything, they will be required to sell to the highest bidder all real property, including shares in corporations, that passes to them from people's estates. Administering these sales will be a major and costly burden. Yet even today, with inheritance, much of the property from most estates where there is no surviving spouse is put up for sale. But these sales are conducted privately by the heirs of the estates, the administrative burden thus being shared by many, rather than falling upon the government. Of course, if inheritance is abolished, not all of the burden of selling property from people's estates will fall upon the government either; numerous charitable organizations will, together, shoulder much, if not most, of the burden.

But the government will probably still have a sizeable amount of property to dispose of. This administrative burden, along with the additional burden of administering any methods for stimulating savings and investments that may become necessary, qualifies as a genuine disadvantage of abolishing inheritance.

7. Finally, abolishing inheritance might be objected to on the grounds that it deprives people of property that clearly has more value for them than for others. Take, for example, precious family heirlooms. Although their monetary value for those within and outside the family is the same, for those within the family they have, in addition to their monetary value, a sentimental value as well. Abolishing inheritance, it might be argued, deprives families of these precious heirlooms, allowing them to pass to others for whom they have far less value. But, more serious yet, consider the son or daughter who has taken over the family business or farm, and who has neither the training nor desire to pursue any other occupation. Again, for such a person the business or farm can be expected to have value far above the mere monetary value it has for others; indeed, he or she may have built a whole life around it. Yet abolishing inheritance will take the business or farm away from this person and allow it to pass to someone for whom it has far less value.

The family heirloom problem seems less serious than that of the family business or farm. According to the proposal for abolishing inheritance we are considering, if the decedent had provided for it in his will, family members would have a chance to buy any family heirlooms before they went on the open market and, except for the most expensive heirlooms, they would normally have the means to do so. If the decedent had provided for it, a son or daughter would be given the same chance to buy a family business or farm. But even with rather liberal credit terms, probably most sons and daughters would not be in a strong enough financial position to take advantage of this opportunity.

Yet it is important to realize that even the family business/farm problem is, today, rather limited in scope. There was a period in history when disrupting occupational continuity from parent to child by abolishing inheritance would have had grave consequences for a nation as a whole. Higher education and other occupational training were not readily available for most people; more often than not, children learnt their occupation from their parents so as to carry on with the family business or farm after their parents died. To disrupt this continuity by abolishing inheritance, thereby depriving vast numbers of people of the only job for which they were adequately prepared, might well have been a national economic disaster. Inheritance, in those days, might well have been one of society's main devices for achieving continuity from one generation to the next in the production of goods and services.

But those days have long since passed—and, incidentally, I see little reason for trying to resurrect them. Today higher education and other forms of occupational training are much more available; people are far more mobile than before; children do not, for the most part, follow in the exact footsteps of their parents anymore; and continuity from one generation to the next in the production of goods and services is now achieved instead largely through the separation of ownership from management that is found in today's large corporations. Disrupting occupational continuity from parent to child by abolishing inheritance might have serious consequences for the limited number of children who would like to carry on from their parents, but the number is indeed limited.

It might be pointed out, however, that the objection to depriving children of family businesses or farms goes beyond just the bad consequences of doing so. This objection, so it might be said, also involves a matter of justice; after a child has already built his or her life around a family business or farm on the understanding that he or she will be able to inherit it someday, for society then to change its mind, abolish inheritance and take the business or farm away, would be unjust. I agree. But what this proves is not that abolishing inheritance is unjust, but that if it is abolished, its abolishment should not be applicable to those who have already built their lives upon the understanding that property could be inherited; rather, its abolishment should be applicable only to those not yet born or too young to have been thus misled. The best way is to abolish inheritance gradually over a number of years so as to avoid any frantic rush to have children before the 'deadline', and to avoid the injustice of a child's being able to inherit without restriction while his younger sibling, born after the 'deadline', can inherit nothing. In sum, although the potential loss of family heirlooms and, even more so, of family businesses and farms does seem to be another genuine disadvantage of abolishing inheritance, this disadvantage does not appear to be overwhelming.

It is time now to take stock of where we stand. In the previous section we found that inheritance is inconsistent with the ideals of distribution according to productivity, equal opportunity, and freedom. Thus we began this section with a presumption against inheritance being justified. But this was only a presumption; it could be overridden if we succeeded in finding objections to abolishing inheritance that were strong enough to outweigh this presumption. In this section we have been examining the most common and most serious potential objections. Most of them, however, turned out, upon closer scrutiny, to be largely unsubstantiated. The two objections that clearly could be substantiated were the one concerning administrative costs and the one concerning family heirlooms, businesses, and farms. It remains only to ask whether these two objections are strong enough to tip the balance back in favour of inheritance.

It is impossible to determine exactly how great the administrative costs of abolishing inheritance would be. This, as we saw, depended on the extent to which government would have to be involved in the sale of property from estates and the extent to which it would have to encourage investments, two matters that we cannot know in advance There was some reason to believe that administrative costs might not be extensive. But even if they were extensive, surely they would have to be far more so than it is reasonable to believe they ever would be to override the presumption against inheritance. For this presumption rests upon considerations of justice and freedom against which mere administrative costs should have, I would think, relatively little weight.

But what about the second objection to abolishing inheritance: that it will deprive families of family heirlooms and, more seriously, family businesses and farms? In weighing the importance of this objection we need to keep in mind the main rationale behind it: by abolishing inheritance, we prevent people within the decedent's family from getting what would have been of far more value to them (for sentimental reasons, etc.) than to those who do get it. No doubt abolishing inheritance will redistribute wealth so that some of its value is, in this way, lost. But to offset this loss, there will be a significant gain in the value of this wealth, a gain attributable to the fact that wealth has diminishing marginal utility and that, therefore, if it is distributed more evenly by means of abolishing inheritance, this will (other things being equal) increase its overall utility, or value (see Sect. 3.5 above). Indeed, this gain in its value should more than compensate for any loss attributable to family heirlooms, businesses, and farms—especially since, as we saw, any such loss will be far less than it would have been during an earlier period in history. So we must conclude, I think, that the family heirloom/business/farm objection and the administrative cost objection are not, neither by themselves nor combined, weighty enough to override our initial presumption against inheritance.

And even these two genuine disadvantages of abolishing inheritance are, for the most part, avoided by the compromise that I shall propose next.

6.5 A COMPROMISE PROPOSAL

It is not people getting something for nothing that is objectionable about inheritance; it is that some get it and some don't. In other words, it is not inheritance *per se* that is objectionable, it is the enormously unequal manner in which inherited wealth is distributed, along with the resulting inequalities of opportunity and immense concentrations of economic power, that is objectionable. So, rather than abolishing inheritance altogether, what I am proposing, finally, is this: let us impose no limits at all upon the amount a person is allowed to bequeath or give away; indeed, let us not even impose any taxes upon what a person bequeaths or gives away (the government can raise whatever it needs through other forms of taxation); but let us impose a strict limit upon the amount any one person, in a lifetime, is allowed to receive; let us, in other words, establish a lifetime inheritance quota. (This could, more formally, be called a lifetime 'assessions' quota, but the more common term 'inheritance' is perhaps preferable, provided it is understood that here 'inheritance' includes not just bequests, but gifts as well.)

Aside from the quota itself, this proposal is much the same as the proposal for abolishing inheritance that I outlined at the beginning of the section. The proposal for abolishing inheritance presupposes some criterion for distinguishing between gifts that are, and ones that are not, prohibited. I did not attempt to work out the best criterion for doing this, but whatever it is, this criterion is to be part of this quota proposal as well, its function as part of the quota proposal being to distinguish between gifts that do, and ones that do not, count towards the fulfillment of the quota. And the three exceptions to the abolishment of inheritance—an unlimited exception for one's spouse, an unlimited exception for charitable organizations, and a limited exception for orphaned children and other dependents—are to be applicable to the quota proposal also, but as exceptions to the quota on inheritance rather than to its abolishment.

Some of the advantages of an inheritance quota might be achieved as well with a progressive inheritance (as opposed to estate) tax. None the less, an inheritance quota is superior. For one thing, a quota is simpler and more straightforward. Although, with an inheritance quota, a person receiving a gift or bequest will have to report it to the government so that the government can make sure the quota has not been exceeded, no complex tax calculations ever have to be made, and no tax money ever has to change hands. Probably the least burdensome way to handle the reporting of gifts and bequests is simply by requiring that people supplement their annual report to the government of yearly income with a report of gifts and bequests received during the year. And, to help prevent people from inadvertently exceeding the quota, the government could include, along with the tax forms they already send us, a statement disclosing exactly how close we were, as of the beginning of the year, to having reached the quota. (Anyone who did exceed his quota is to be liable to the government for the excess, plus enough interest so as to make exceeding the quota an unattractive means of, in effect, securing a short-term loan.) By thus combining the administration of the inheritance quota with the administration of income taxes, the administrative burden for both government and individuals will probably be less than it would be with an inheritance tax.

Moreover, there just seems to be something about an inheritance tax that invites unscrupulous politicians, sooner or later, to build insidious loopholes of every conceivable sort into it. But the most important advantage of a quota over a tax is that a quota, but not a tax, does indeed place a ceiling upon how much any one person can receive from all sources throughout his or her life. The absence of such a ceiling is the most insidious loophole of all.

To help potential donors find people who had not yet received all, or any, of their quota, how much everyone had received to date could be made a matter of public record, accessible by anyone, much as are the titles to land today. The beneficiaries of a person's will might always, of course, fulfill their quotas before that person died, thus rendering the bequest void. But this poses no special problems, much less any 'mind-boggling' ones as John Elster has suggested (1986: 147 n. 7). A person could always, in his will, name a number of contingent beneficiaries, in some order of priority, who would inherit to the extent, and only to the extent, that the initial beneficiaries had already received their quotas.

But the crux of any quota proposal is, of course, the amount at which the quota is set. Consider, for example, the application of this quota in the United States for 1990. Any fixed amount, say $50,000, would, due to changes in the value of the dollar, soon become outmoded. So, rather than some fixed amount, it would be wise to set the quota so that, according to some predetermined formula, it would vary in amount according to changing conditions over the years. On the other hand, this formula should be such that it would not allow the quota to vary in amount too often. In view of these considerations, I tentatively suggest that the quota be set at an amount equal to the average dollar value (rounded off to the nearest ten thousand) of all the estates of all those who, at age twenty-one or over, died during the five years previous to the year in which the quota is to be applicable. I say 'tentatively', since it is not clear exactly what this amount currently is, or how a lifetime inheritance quota might affect it. Instead of determining the quota by reference to the average estate, a possible alternative is to determine it by reference to the average net worth of people, age twenty-one or over. Because data on the distribution of wealth in the United States, and on the real value of estates (as opposed to their value merely for estate tax purposes) are surprisingly unavailable, we cannot be certain what average estate and net worth figures are these days. We may gain a rough idea, however, from a 1983 government study, that places the average net worth of Americans in 1983 at $66,050 (Avery *et al.* 1984: 683 n. 2). Let us simply assume, for purposes of argument, that, for 1990, a quota determinable by reference to average estate value, or to average net worth, would be $100,000. (If after a certain number of years this quota went up, say, 10 per cent, then whatever amount a person was still eligible to receive would go up 10 per cent as well; the person who was no longer eligible to receive anything would be unaffected by any increase.)

A quota of $100,000, for 1990, would be modest enough, I think, so that wealthy people, rather than being able to leave each of a few people a large amount, would be forced instead to leave each of a lot of people a smaller amount; so probably far more people would receive a significant amount of inheritance than do today. And since this quota would in fact be a relatively modest one, it would succeed in breaking up the enormous concentrations of wealth that, today, can be passed on from generation to generation. And if enormous concentrations of wealth could not be passed on, those who received little or nothing would not, thereby, be at an enormous competitive disadvantage, as they are today.

On the other hand, this quota would be large enough, I think, so as to avoid, for the most part, the administrative costs of abolishing inheritance. Relatively few people would have estates larger than, say, five times the quota, and since most people would know at least five eligible people to whom they would want to leave their wealth (children, grandchildren, nephews, nieces, close friends, loyal employees, and the like), relatively few estates would therefore require any more from the government than an appraisal of assets, something that the government already does under current estate and gift tax law anyway. And many people would have a motive for saving and investment they would not have if inheritance were abolished altogether—this motive being that of leaving wealth to these individuals. Finally, this quota would be large enough to significantly alleviate the family heirloom/business/farm problem as well.

One final point: if, because of a modest lifetime inheritance quota, a family business became too valuable to be bequeathed to family members, then this opens up the possibility of its being purchased by the employees. Indeed, the government, through appropriate legislation, should encourage purchases by employees. The government might, for example, pass legislation that guaranteed employees favourable credit terms for any such purchases. What I have in mind here is the sort of legislation proposed in Chapter 4. As explained in some detail there, in return for favourable credit terms, the government would require that the business take on a 'worker-control' format according to which, among other things, all future, full-time employees are to purchase a share in the business as well. Moreover, this legislation should also allow for bequests of businesses to workers on the condition that the business take on a 'worker-control' format. (Any workers whose quotas would be exceeded by their share of such a bequest would, of course, be given the option of 'buying-in' with deferred payments, as explained in Ch. 4.) To further encourage such bequests, this legislation might specify that the testator, if a majority owner, has the right to stipulate what name the business shall have for as long as it continues to exist. The testator could thereby choose to have the business carry on with his name, so that it might become a kind of memorial to him or his family.

But we need not try to work out the details of this legislation here. The general point is that a modest lifetime inheritance quota, combined with the right sort of supportive legislation, might do much to promote worker control. The potential gains from worker control are considerable: they include less worker alienation, work that is more satisfying, and a greater share for workers in the profits from their own work (see Ch. 4). If a modest lifetime inheritance quota, combined with the right sort of supportive legislation, did succeed in promoting a gradual, and voluntary, evolution towards worker control, this would be a significant bonus indeed.

6.6 CAPITALISM WITH MORALITY: CONCLUSION

I have argued, in the final three chapters of this book, for an economic system with the following seven features.

1. Private ownership of the means of production and, in general, freedom from central economic planning.

2. A lifetime inheritance quota.

3. Measures promoting worker-controlled enterprises that include legislative recognition of the worker-control format and favourable credit terms for enterprises that have, or are willing to adopt, this format.

4. A refundable, earned income credit (EIC) that is sufficient to assure virtually all working families of a family income above the poverty line. This is to be combined with other tax reforms that include a standard exemption equal to the poverty line for an individual with no dependents, and the elimination of many middle- or upper-class tax deductions and of all tax benefits for special interests.

5. Income-related, cost-sharing government programmes sufficient for assuring that virtually everyone can afford decent medical care, college or occupation training, and preschool child care. These programmes are to try to make medical, educational, and child-care expenses more or less 'income neutral'.

6. Measures for assuring full employment. These measures are to include abolishing minimum wages, and (as called for by feature 5) subsidizing educational and child-care expenses. They may include government tax breaks for employee income in the form of a fixed percentage share in the enterprise's revenue, and, as a last resort, government-created jobs.

7. Measures for stimulating capital accumulation and thus economic growth—in so far as needed and compatible with preserving the environment. These are to include measures for stimulating not just physical capital, but human and knowledge capital as well, measures such as making college or occupational training fully available to everyone at all income levels (feature 5).

These seven features are closely interrelated in that the effectiveness of each depends largely upon implementing the others. Taken together, they delineate an economic system. I am not sure which of these features may be said to be part of the economic system itself, thus qualifying them as defining features, and which are instead extraneous to the system, but designed to compensate for its deficiencies. It does not matter, however, since this system is to be viewed as incorporating all these features, whether they be 'defining' ones or not. The resulting system is a form of capitalism, since the means of production are to be privately owned and, in general, the market is to be free from central planning. Yet it is a form of capitalism that differs significantly from any current forms. It is designed so as to encourage worker-controlled enterprises to coexist along with traditional ones. And it is designed to achieve a more complete realization that does any current system of four crucial ideals of political morality: equal access to the basic necessities of life; equal opportunity; productivity; and freedom.

This system achieves equal access to necessities through an earned income credit (feature 4), through government subsidies of medical care, child care, college, and occupational training (feature 5), through measures helping to assure that there will be a decent job available for anyone who wants one (feature 6), and, of course, through unemployment, disability, and retirement benefits such as those already found in large, industrialized countries. More generally, however, this system achieves equal access to necessities by dispersing wealth more widely throughout society than does any form of capitalism today. This wide dispersement of wealth is achieved, first and foremost, through a lifetime inheritance quota (feature 2). And, to the extent that the measures for promoting worker control are successful (feature 3), wealth will be dispersed even more widely since worker control spreads investment wealth among workers rather than allowing it to concentrate largely among just a relatively few, exceptionally rich individuals.

The very fact that, with this system, wealth will be more widely dispersed does more than perhaps anything else could to promote equal opportunity. And, incidentally, equal opportunity promotes a more wide dispersement of wealth; the two values are mutually reinforcing. Equal opportunity is promoted to a great extent as well by the governmental measures of this system that make decent medical care, child care, college and occupational training fully available to everyone at all income levels (feature 5). Also, with more worker-controlled enterprises comes more opportunity for workers to speak out, influence company decisions, and even occupy managerial positions.

The equal opportunity advanced by this system promotes economic productivity. The more equal within a system that opportunity is, the more close the correspondence will be between people's occupations and their qualifications; and the more close the correspondence between occupations and qualifications, the greater the productivity. Other means by which this system promotes productivity include measures that promote full employment (feature 6), and that promote economic growth (feature 7). In general, this system promotes productivity, for reasons explained in Section 3.6, by its fundamental commitment to a free market (feature 1) and thus to eliminating any governmental measures that notably impede the market's effectiveness, measures such as minimum wages, tax benefits for special interests, and poorly conceived welfare programmes.

This commitment to the free market also, for reasons explained in Section 3.5, helps promote important dimensions of freedom: entrepreneurial freedom; consumer freedom; and political freedom, including freedom of speech. The income-related, cost-sharing of educational and medical expenses (feature 5) helps promote freedom from the constraints of ignorance and disease. The measures for promoting wide dispersement of wealth, equal access to basic necessities, and productivity all serve to promote freedom from the constraints of poverty. And, to the extent that these various measures succeed in eliminating poverty and dispersing wealth more widely, they succeed also in breaking down class barriers, thus promoting 'social' freedom and self-respect. Moreover, in so far as political power in a democracy is a function of wealth, these same measures serve to promote political freedom as well. Finally, the measures for promoting worker control (feature 3) promote freedom from the constraints of a totalitarian workplace.

As should be apparent from this discussion, these four interrelated ideals—equal access to necessities, equal opportunity, productivity, and freedom—are hardly incompatible with one another. On the contrary, they very much reinforce one another. Nevertheless, the economic system proposed here does not realize any one of these ideals, taken by itself, to the fullest extent possible. Instead, this system is put forth as the best possible compromise, a compromise that manages to promote each ideal without seriously undermining any of the others, or any other important values. This system, in other words, is put forth as the one most in the general welfare. And, of course, all of this presupposes that account of the general welfare, and of morality, set out in the first chapter and Section 2.2, an account designed to avoid the dead end of scepticism.

As I said at the beginning of this book, the critics of capitalism see the extreme inequalities of wealth and opportunity that it breeds, the burdens it places upon ordinary working people, and they conclude that capitalism is immoral. I have argued that these critics are wrong. It is not capitalism *per se* that is immoral, but *current* capitalism. Capitalism with morality is possible. Throughout this book I have tried to suggest what the main features of a new, more humane capitalism might be. These suggestions are, of course, only tentative; they can be improved. Yet they may help us see that a new, more humane capitalism is indeed possible, a capitalism without extreme inequalities of wealth and opportunity, a capitalism without alienated workers, a capitalism with morality.

REFERENCES

AARON, HENRY, and SCHWARTZ, WILLIAM B. (1990), 'Rationing Health Care: The Choice before Us', *Science*, 247: 418–22.

ALCHIAN, ARMEN A., and DEMSETZ, HAROLD (1972), 'Production, Information Costs, and Economic Origination', *American Economic Review*, 62: 777–95.

ARNOLD, N. SCOTT (1989), 'Marx, Central Planning, and Utopian Socialism', in Ellen Frankel Paul, Fred D. Miller, Jun., and Jeffrey Paul (eds.), *Socialism* (Oxford: Blackwell), 160–90.

—— (1990), *Marx's Radical Critique of Capitalist Society* (Oxford: Oxford University Press).

AVERY, R., ELLIEHOUSEN, G., CANNER, G., and GUSTAFSON, T. (1984), 'Survey of Consumer Finances, 1983: Second Report', *Federal Reserve Bulletin*, 70 (Dec.).

Background Material and Data on Programs within the Jurisdiction of the Committee on Ways and Means (1989) (Washington, DC: Government Printing Office).

BARLOW, ROBIN, BRAZER, HARVEY E., and MORGAN, JAMES N. (1966), *Economic Behavior of the Affluent* (Washington, DC: Brookings Institution).

BECK, R. G. (1974), 'The Effects of Co-Payment on the Poor', *Journal of Human Resources*, 9: 128–42.

BENTHAM, JEREMY (1789), *The Principles of Morals and Legislation*, many editions.

BLINDER, ALAN S. (1987), *Hard Heads, Soft Hearts* (Reading, Mass.: Addison-Wesley).

BRANDT, RICHARD B. (1979), *A Theory of the Good and Right* (Oxford: Clarendon Press).

—— (1982), 'Two Concepts of Utility', in Harlan B. Miller and William H. Williams (eds.), *The Limits of Utilitarianism* (Minneapolis: University of Minnesota Press), 169–85.

BRITTAIN, JOHN A. (1978), *Inheritance and the Inequality of National Wealth* (Washington, DC: Brookings Institute).

BROOME, J. A. (1978), 'Choice and Value in Economics', *Oxford Economic Papers*, 30: 213–33.

BROWNING, EDGAR K. (1975), *Redistribution and the Welfare System* (Washington, DC: American Enterprise Institute).

BUCHANAN, ALLEN (1985), *Ethics, Efficiency, and the Market* (Totowa, NJ: Rowman & Allanheld).

CARENS, JOSEPH H. (1981), *Equality, Moral Incentives, and the Market* (Chicago: University of Chicago Press).

CHAMBERS, DONALD (1981), 'Another Look at Poverty Lines in England and the United States', *Social Service Review*, 55 (Sept.).

CHRISTIE, DREW (1984), 'Recent Call for Economic Democracy', *Ethics*, 95: 112–28 (Oct.).

CLAYRE, ALASDAIR (1980), 'Some Aspects of the Mondragon Cooperation Federation', in Alasdair Clayre (ed.), *The Political Economy of Co-operation and Participation* (Oxford: Oxford University Press), 171–4.

COHEN, G. A. (1978), 'Labor, Leisure, and the Distinctive Contradiction of Advanced Capitalism', in Gerald Dworkin, Gordon Bermant, and Peter G. Brown (eds.), *Market and Morals* (Washington, DC: Hemisphere), 107–36.

COHEN, JOSHUA, and ROGERS, JOEL (1983), *On Democracy: Toward a Transformation of American Society* (New York: Penguin Books).

COOPER, GEORGE A. (1979), *A Voluntary Tax? New Perspectives on Sophisticated Estate Tax Avoidance* (Washington, DC: Brookings Institution).

DEWEY, JOHN (1939), *Theory of Valuation* (Chicago: University of Chicago Press).

DOWIE, MARK (1977), 'How Ford Put Two Million Firetraps on Wheels', *Business and Society Review*, 23: 46–55 (Fall).

DUNCAN, GREG J., and HOFFMAN, SAUL D. (1988), 'The Use and Effects of Welfare: A Survey of Recent Evidence', *Social Service Review* (June).

DWORKIN, RONALD (1978), *Taking Rights Seriously* (Cambridge, Mass.: Harvard University Press).

ELSTER, JON (1986), 'Comments on Krouse and McPherson', *Ethics*, 97: 146–53.

—— (1989), 'From Here to There', in Ellen Frankel Paul, Fred D. Miller, Jun., and Jeffrey Paul (eds.), (Oxford: Blackwell), 93–111.

—— and MOENE, KARL OVE (1989), *Alternatives to Capitalism* (Cambridge: Cambridge University Press).

ESTRIN, SAUL (1989), 'Workers' Co-operatives: Their Merits and their Limitations', in Julian Le Grand and Saul Estrin (eds.), *Market Socialism* (Oxford: Clarendon Press).

EVERS, WILLIAMSON M. (1989), 'Liberty of the Press under Socialism', in Ellen Frankel Paul, Fred D. Miller, Jun., and Jeffrey Paul (eds.), *Socialism* (Oxford: Blackwell), 211–34.

FEINBERG, JOEL (1973), *Social Philosophy* (Englewood Cliffs, NJ: Prentice-Hall).

—— (1980), 'The Idea of a Free Man', *Rights, Justice, and the Bounds of Liberty* (Princeton, NJ: Princeton University Press), 3–29.

FELDSTEIN, M. S. (1971), 'A New Approach to National Health Insurance', *Public Interest* (Spring).

FIEKOWSKY, SEMOUR (1959), 'On the Economic Effects of Taxation in the United States', unpublished dissertation (Harvard University).

FRIEDMAN, DAVID (1973), *The Machinery of Freedom* (New York: Harper & Row).

FRIEDMAN, MILTON (1962), *Capitalism and Freedom* (Chicago: University of Chicago Press).

—— and FRIEDMAN, ROSE (1980), *Free to Choose* (New York: Harcourt Brace Jovanovich).

GALBRAITH, JOHN KENNETH (1967), *The New Industrial State* (Boston: Houghton Mifflin).

GAUTHIER, DAVID (1986), *Morals by Agreement* (Oxford: Clarendon Press).

GLAZER, NATHAN (1986), 'Education and Training Programs and Poverty', in Sheldon H. Danziger and Daniel H. Weinberg (eds.), *Fighting Poverty* (Cambridge, Mass.: Harvard University Press), 152–72.

GLOVER, JONATHAN (1977), *Causing Death and Saving Lives* (Harmondsworth: Penguin).

GRAY, JOHN (1989), 'Against Cohen on Proletarian Unfreedom', in Ellen Frankel Paul, Fred D. Miller, Jun., Jeffrey Paul, and John Ahrens (eds.), *Capitalism* (Oxford: Blackwell), 77–112.

GREENWOOD, DAPHNE (1973), 'An Estimation of U.S. Family Wealth and its Distribution from Micro Data, 1973', *Review of Income and Wealth*, 26–66.

GRICE, RUSSELL (1967), *The Grounds of Moral Judgment* (Cambridge: Cambridge University Press).

GRIFFIN, JAMES (1986), *Well-Being* (Oxford: Clarendon Press).

HALPER, THOMAS (1991), 'Rights, Reforms, and the Health Care Crisis: Problems and Prospects', in Thomas J. Bole, III, and William B. Bondeson (eds.), *Rights to Health Care* (Dordrecht: Kluwer Academic Publishers), 135–68.

HANSSON, INGEMAR (1987), 'An Evaluation of the Evidence on the Impact of Taxation of Capital Formation', in Hans Van der Kar and Barbara Wolfe (eds.), *The Relevance of Public Finance* (Detroit: Wayne State University Press), 1–10.

HARE, R. M. (1963), *Freedom and Reason* (Oxford: Oxford University Press).

—— (1981), *Moral Thinking: Its Levels, Method and Point* (Oxford: Clarendon Press).

HARRIS, JOHN (1975), 'The Survival Lottery', *Philosophy*, 50 (Jan.).

HARSANYI, JOHN C. (1982), 'Morality and the Theory of Rational Behavior', in Amartya Sen and Bernard Williams (eds.), (Cambridge: Cambridge University Press), *Utilitarianism and Beyond*, 39–62.

HASLETT, D. W. (1986), 'Is Inheritance Justified?' *Philosophy & Public Affairs*, 15: 122–55 (Spring).

—— (1987a), *Equal Consideration: A Theory of Moral Justification* (Newark, Del.: University of Delaware Press; London and Toronto: Associated University Presses).

—— (1987b), 'What is Wrong with Reflective Equilibria?' *Philosophical Quarterly*, 37: 305–11.

—— (1990). 'What is Utility?' *Economics and Philosophy*, 6: 65–94 (Spring).

HAVEMAN, ROBERT (1988), *Starting Even* (New York: Simon & Schuster).

HAYEK, F. A. (1935) (ed.), *Collectivist Economic Planning* (London: Routledge & Kegan Paul).

—— (1960), *The Constitution of Liberty* (Chicago: University of Chicago Press).

—— (1967), *Collectivist Economic Planning* (New York: Kelley).

HEILBRONER, ROBERT L., and THUROW, LESTER (1975), *Understanding Microeconomics*, 3rd edn. (Englewood Cliffs, NJ: Prentice-Hall).

—— (1989), 'The Triumph of Capitalism', *New Yorker*, 64: 98–109.

HIRSHLEIFER, JACK (1976), *Price Theory and Applications* (Englewood Cliffs, NJ: Prentice-Hall).

HOOVER, G. E. (1927), 'The Economic Effects of Inheritance Taxes', *American Economic Review*, 17: 38–49.

HUTCHINSON, HARRY D. (1973), *Economics and Social Goals: An Introduction* (Chicago: Science Research Associates, Inc.).

268 References

JAY, PETER (1980), 'The Workers' Co-operative Economy', in Alasdair Clayre (ed.), *The Political Economy of Co-operation and Participation* (Oxford: Oxford University Press), 9–45.

JENSEN, MICHAEL, and MECKLING, WILLIAM (1979), 'Rights and Production Functions: An Application to Labor-Managed Firms and Codetermination', *Journal of Business*, 52: 469–506.

JOHNSON, ALVIN H. (1914), 'Public Capitalization of the Inheritance Tax', *Journal of Political Economy*, 22: 160–80.

KAHN, ROBERT L. (1972), 'The Meaning of Work: Interpretation and Prospects for Measurement', in A. A. Campbell and P. E. Converse (eds.), *Human Meaning and Social Change* (New York: Russell Sage Foundation), 159–203.

KASL, STANISLAV (1977), 'Work and Mental Health', in W. T. Heisler and John W. Houck (eds.), *A Matter of Dignity* (Notre Dame, Ind.: University of Notre Dame Press), 85–110.

KAUS, MICKEY (1992), *The End of Equality* (New York: Basic Books).

KIRKLAND, FRANK M. (1992), 'Social Policy, Ethical Life, and the Urban Underclass', in Bill E. Lawson (ed.), *The Underclass Question* (Philadelphia: Temple University Press).

KORNHAUSER, A. W. (1965), *Mental Health of the Industrial Worker: A Detroit Study* (Huntington, NY: R. E. Krieger).

KROUSE, RICHARD, and McPHERSON, MICHAEL (1986), 'A Mixed-Property Regime: Equality and Liberty in a Market Economy', *Ethics*, 97: 119–38.

LANGE, OSKAR (1964 [1938]), 'On the Economics of Socialism', in B. E. Lippencott (ed.), *On the Economics of Socialism* (New York: McGraw Hill).

LERNER, ABBA P. (1946), *The Economics of Control: Principles of Welfare Economics* (New York: Macmillan).

LETWIN, WILLIAM (1983), *Against Equality* (London: MacMillan).

LEWIS, C. I. (1946), *An Analysis of Knowledge and Valuation* (LaSalle, Ill.: Open Court).

LOCKE, JOHN (1690), *Two Treatises of Government*.

LOUIS, ARTHUR M. (1968), 'America's Antimillionaires', *Fortune* (May).

—— (1973), 'The New Rich of the Seventies', *Fortune* (Sept.).

LYDALL, HAROLD (1984), *Yugoslav Socialism: Theory and Practice* (Oxford: Clarendon Press).

LYONS, DAVID (1977), *Forms and Limits of Utilitarianism* (Oxford: Clarendon Press), originally published 1965.

MACHAN, TIBOR R. (1986), 'The Virtue of Freedom in Capitalism', *Journal of Applied Philosophy*, 3: 49–58.

MARCUSE, HERBERT (1964), *One Dimensional Man* (Boston: Beacon Press).

MARX, KARL (1867), *Capital*, vol. i, many editions.

McCAIN, ROGER A. (1977), 'On the Optimum Financial Environment for Worker Cooperatives', *Zeitschrift für Nationalokonomie*, 37: 355–84.

McCLELLAND, D. C. (1961), *The Achieving Society* (Princeton, NJ: Van Nostrand).

McLAUGHLIN, ANDREW (1972), 'Freedom versus Capitalism', in Tibor R. Machan (ed.), *The Main Debate: Communism versus Capitalism* (New York: Random House), 217–32.

MEADE, JAMES (1980), 'Labour Co-operatives, Participation, and Value-Added

Sharing', in Alasdair Clayre (ed.), *The Political Economy of Co-operation and Participation* (Oxford: Oxford University Press), 89–108.

MILL, JOHN STUART (1859), *On Liberty*, many editions.

—— (1861), *Utilitarianism*, many editions.

—— (1970 [1848]) *Principles of Political Economy* (Harmondsworth: Penguin).

MILLER, DAVID (1989), *Market, State and Community* (Oxford: Clarendon Press).

MISES, LUDWIG VON (1951), *Socialism: An Economic and Sociological Analysis* (New Haven, Conn.: Yale University Press).

MURRAY, CHARLES (1984), *Losing Ground: American Social Policy, 1950–1980* (New York: Basic Books).

NARVESON, JAN F. (1967), *Morality and Utility* (Baltimore: Johns Hopkins Press).

—— (1988), *The Libertarian Idea* (Philadelphia: Temple University Press).

NEWHOUSE, J. P., MANNING, W. G., MORRIS, C. M., ORR, L. L., DUAN, N., KEELER, E. B., LEIBOWITZ, A., MARQUIS, K. H., MARQUIS, M. S., PHELPS, C. E., and BROOK, R. H. (1981), 'Some Interim Results from a Controlled Trial of Cost Sharing in Health Insurance', *New England Journal of Medicine*, 305: 1501–7.

NIELSEN, KAI (1985), *Equality and Liberty* (Totowa, NJ: Rowman & Allanheld).

NORDHAUS, WILLIAM, and TOBIN, JAMES (1972), 'Is Growth Obsolete?' in *Fiftieth Anniversary Colloquium V* (New York: National Bureau of Economic Research, Columbia University Press).

NORMAN, RICHARD (1982), 'Does Equality Destroy Liberty?' in Keith Graham (ed.), *Contemporary Political Philosophy* (Cambridge: Cambridge University Press), 83–109.

NOVE, ALEC (1962 [1958]), 'The Problem of "Success Indicators" in Soviet Industry', *Economica*, 97: 1–13, reprinted in F. Holzman (ed.), *Readings in the Soviet Economy* (Chicago: Rand McNally).

NOZICK, ROBERT (1974), *Anarchy, State, and Utopia* (Oxford: Blackwell).

—— (1989), *The Examined Life: Philosophical Meditations* (New York: Simon & Schuster).

O'NEILL, JOHN (1989), 'Markets, Socialism and Information: A Reformulation of a Marxian Objection to the Market', in Ellen Frankel Paul, Fred D. Miller, Jun. and Jeffrey Paul (eds.), *Socialism* (Oxford: Blackwell), 200–10.

OSBORN, ALBERT F., and MILBANK, JANE E. (1987), *The Effects of Early Education: A Report from the Child Health and Education Study* (Oxford: Oxford University Press).

PEARCE, DAVID W. (1986) (ed.), *The MIT Dictionary of Modern Economics*, 3rd edn. (Cambridge, Mass.: MIT Press).

PECHMAN, JOSEPH A. (1984), 'Fixing the Income Tax with the Fair Tax', in Joseph A. Pechman, Bradley, Bill, Gephardt, Richard, Bell, Joseph, and Heller, Walter, *Tax Policy: New Directives and Possibilities* (Washington, DC: Center for National Policy), 19–24.

PROJECTOR, DOROTHY S. (1964), 'Survey of Financial Characteristics of Consumers', *Federal Reserve Bulletin*, 50 (Mar.).

PUTTERMAN, LOUIS (1982), 'Some Behavioral Perspectives on the Dominance of Hierarchical over Democratic Forms of Enterprise', *Journal of Economic Behavior and Organization*, 3: 139–60.

PUTTERMAN, LOUIS (1984), 'On Some Recent Explanations of Why Capital Hires Labor', *Economic Inquiry*, 22: 171–87.

—— (1990), *Division of Labor and Welfare: An Introduction to Economic Systems* (Oxford: Oxford University Press).

RAWLS, JOHN (1971), *A Theory of Justice* (Cambridge, Mass.: Harvard University Press).

REICH, ROBERT B. (1989), *The Resurgent Liberal* (New York: Times Books).

REINHARDT, UWE (1987), 'Health Maintenance Organizations in the United States: Recent Developments', Princeton University, Mimeograph.

ROBINSON, JOAN (1962), *Economic Philosophy* (Chicago: Aldine).

ROSENBAUM, S. (1983), 'The Prevention of Infant Mortality: The Unfulfilled Promise of Federal Health Programs for the Poor', *Clearinghouse Review*, 17: 701–33.

SARTORIOUS, ROLF E. (1975), *Individual Conduct and Social Norms* (Encino, Calif.: Dickenson).

SCHERER, F. M. (1980), *Industrial Market Structure and Economic Performance* (Chicago: Rand McNally).

SCHWARTZ, THOMAS (1982), 'Human Welfare: What it is not', in Harlan B. Miller and William H. Williams (eds.), *The Limits of Utilitarianism* (Minneapolis: University of Minnesota Press), 195–206.

SCHWEICKART, DAVID (1980), *Capitalism or Worker Control?* (New York: Praeger).

SCHITOVSKY, A., and McCALL, N. (1977), 'Coinsurance and the Demand for Physicians' Services: Four Years Later', *Social Security Bulletin*, 40: 19–27.

SEIDMAN, LAURENCE S. (1979), 'Health Inflation: A Diagnosis and Prescription', *Challenge* (July–Aug.).

—— (1990), *Saving for America's Future* (Armonk, NY: M. E. Sharpe).

SEN, AMARTYA K. (1977), 'Rational Fools', *Philosophy and Public Affairs*, 6: 317–44.

SHAPIRO, DANIEL (1989), 'Reviving the Socialist Calculation Debate: A Defense of Hayek Against Lange', in Ellen Frankel Paul, Fred D. Miller, Jun., and Jeffrey Paul (eds.), *Socialism* (Oxford: Blackwell), 139–59.

SIDGWICK, HENRY (1907), *The Method of Ethics*, 7th edn. (London: Macmillan).

SINGER, PETER (1972), 'Famine, Affluence and Morality', *Philosophy and Public Affairs*, 7: 229–43.

—— (1977), *Animal Liberation* (New York: Cape).

—— (1979), *Practical Ethics* (Cambridge: Cambridge University Press).

SMART, J. J. C. (1973), 'An Outline of a System of Utilitarian Ethics', in J. J. C. Smart and Bernard Williams, *Utilitarianism: For and Against* (Cambridge: Cambridge University Press), 3–74.

SMITH, ADAM (1776), *An Inquiry into the Nature and Causes of the Wealth of Nations*, many editions.

SMITH, RICHARD A. (1957), 'The Fifty-Million Dollar Man', *Fortune* (Nov.).

SUMNER, L. W. (1981), *Abortion and Moral Theory* (Princeton, NJ: Princeton University Press).

—— (1987), *The Moral Foundation of Rights* (Oxford: Clarendon Press).

THUROW, LESTER C. (1971), *The Impact of Taxes on the American Family* (New York: Praeger).

—— (1976), 'Tax Wealth, Not Income', *New York Times Magazine*, 11 (Apr.).

—— (1985), *Zero Sum Solution* (New York: Simon & Schuster).

—— (1992), *Head to Head* (New York: Morrow).

TROTSKY, LEON (1937), *The Revolution Betrayed* (New York: Pathfinder Books).

TULLOCK, GORDON (1971), 'Inheritance Justified', *Journal of Law and Economics* (Oct.).

US Bureau of the Census (1990), Series P-60, No. 169-RD (Washington, DC: US Government Printing Office).

VANEK, JAROSLAV (1977), *The Labor-Managed Economy* (Ithaca, NY: Cornell University Press).

WALDMAN, MICHAEL (1990), *Who Robbed America?* (New York: Random House).

WALZER, MICHAEL (1983), *Spheres of Justice* (New York: Basic Books).

WARD, BENJAMIN (1967), *The Socialist Economy: A Study of Organizational Alternatives* (New York: Random House).

WARNOCK, G. J. (1971), *The Object of Morality* (London: Methuen).

WATERS, JAMES A. (1989), 'Corporate Morality as an Organizational Phenomenon', in A. Pablo Iannone (ed.), *Contemporary Moral Controversies in Business* (Oxford: Oxford University Press), 151–63.

WEITZMAN, MARTIN L. (1983), 'Some Macroeconomic Implications of Alternative Compensation Systems', *Economic Journal*, 93 (Dec.).

—— (1984), *The Share Economy* (Cambridge, Mass.: Harvard University Press).

WILLIAMS, BERNARD (1962), 'The Idea of Equality', in Peter Laslett and W. G. Runciman (eds.), *Philosophy, Politics and Society*, ser. 2 (Oxford: Basil Blackwell), 110–31.

WILLIAMSON, OLIVER E. (1975), *Markets and Hierarchies: Analysis and Antitrust Implications* (New York: Free Press).

—— (1980), 'The Organization of Work: A Comparative Institutional Assessment', *Journal of Economic Behaviour and Organization*, 1 (Mar.).

WILSON, WILLIAM JULIUS (1987), *The Truly Disadvantaged* (Chicago: University of Chicago Press).

YOUNG, MICHAEL (1961), *The Rise of Meritocracy* (Harmondsworth: Penguin).

INDEX